Urban Sustainability

The Urban Sustainability Book Series is a valuable resource for sustainability and urban-related education and research. It offers an inter-disciplinary platform covering all four areas of practice, policy, education, research, and their nexus. The publications in this series are related to critical areas of sustainability, urban studies, planning, and urban geography.

This book series aims to put together cutting-edge research findings linked to the overarching field of urban sustainability. The scope and nature of the topic are broad and interdisciplinary and bring together various associated disciplines from sustainable development, environmental sciences, urbanism, etc. With many advanced research findings in the field, there is a need to put together various discussions and contributions on specific sustainability fields, covering a good range of topics on sustainable development, sustainable urbanism, and urban sustainability. Despite the broad range of issues, we note the importance of practical and policy-oriented directions, extending the literature and directions and pathways towards achieving urban sustainability.

The series will appeal to urbanists, geographers, planners, engineers, architects, governmental authorities, policymakers, researchers of all levels, and to all of those interested in a wide-ranging overview of urban sustainability and its associated fields. The series includes monographs and edited volumes, covering a range of topics under the urban sustainability topic, which can also be used for teaching materials.

Ali Cheshmehzangi

Urban Innovation and Sustainability

ICT-based Interventions in Hainan, China

Springer

Ali Cheshmehzangi
School of Architecture, Design
and Planning
The University of Queensland
Brisbane, QLD, Australia

ISSN 2731-6483 ISSN 2731-6491 (electronic)
Urban Sustainability
ISBN 978-981-96-6369-9 ISBN 978-981-96-6370-5 (eBook)
https://doi.org/10.1007/978-981-96-6370-5

This Springer imprint is published by the registered company Springer Nature Singapore Pte Ltd.
The registered company address is: 152 Beach Road, #21-01/04 Gateway East, Singapore 189721, Singapore

Xiaodonghai beach, where I realised there is no fish in the sea! Photo taken in 2021, by Ali Cheshmehzangi

For the exhausted soil of my patria, with its loneliness and forgotten desires.

In remembering a nation:
Purposefully exploited, raged, sabotaged, inactivated, abused, …

Preface

"Hainan will become a paradigm of reform and opening-up in the new era".

—President Xi Jinping, 2022, during his inspection visit to Hainan

In an era where cities and communities are increasingly dependent on technology to address urban challenges, the integration of Information and Communication Technologies (ICTs) has become pivotal in fostering sustainable development. This dependence is not merely a consequence of technological advancement but a fundamental shift in the way societies conceptualize progress, governance, and collective well-being. Cities, as dynamic entities, are constantly evolving in response to economic, environmental, and social imperatives. The ability to harness digital technologies to create sustainable, resilient, and inclusive urban environments reflects a deeper philosophical transformation—one that prioritizes adaptability, efficiency, and human-centric innovation over static models of urban planning.

Technology, in its essence, is a double-edged sword; it holds the promise of enhanced efficiency and connectivity but also presents risks of exclusion, over-reliance, and unintended socio-political consequences. The responsible integration of ICTs requires a vision that extends beyond the immediate gains of digitization—it necessitates a commitment to long-term sustainability, ensuring that technological interventions remain equitable, ethical, and inclusive. Digital tools, when applied thoughtfully, have the power to bridge social gaps, facilitate participatory governance, and democratize access to essential resources. However, their true potential is realized only when they are embedded within a framework that prioritizes social equity, environmental stewardship, and economic viability.

This collection of context-specific ICT-based interventions under the title of "*Urban Innovation and Sustainability*" aims to accomplish sustainability in a number of ways. It acknowledges that sustainability is not a singular goal but a multi-faceted, evolving process that requires continuous adaptation. By examining how ICT applications influence the environmental, social, economic, and institutional dimensions of urban life, this book presents a delicate perspective on the role of technology in shaping sustainable futures. The case studies included here serve as testaments to the transformative power of ICTs in urban (as well as non-urban) contexts, offering insights into their potential to enhance environmental protection, foster social well-being, create economic opportunities, and strengthen governance structures.

Hainan, an island province with aspirational sustainability objectives, has become a test site for creative urban solutions. Five ICT-based pilot studies that examine useful applications in actual settings are highlighted in this book. These studies offer insights into the long-term potential for scalability and policy integration of digital interventions in addition to evaluating their impact and viability. Surveys, case studies, and stakeholder engagement are some of the methods used in this research to document the changing relationship between sustainability and urban innovation.

A wide variety of ICT-based interventions are covered in the case studies included in this book. These initiatives, which range from using technology to protect the environment to improving social well-being and economic opportunities, are prime examples of how digital tools can revolutionize society. The use of ICTs to boost the tourism sector, the role of social media in place promotion, the digitization of local markets to generate online economic opportunities, and health support systems for senior communities are some important areas of focus. These interventions show how digital solutions can be customized to fit community needs, guaranteeing flexibility and inclusivity in a variety of urban environments.

This book recognizes the risks and difficulties of technological interventions while highlighting the benefits of ICT-driven urban transformation. Concerns regarding data privacy, digital divides, and the unexpected effects of automation are brought up by our growing reliance on digital tools. Even though the case studies touch on these issues in passing, the main emphasis is still on how well-executed ICT applications can support paradigm shifts, advance infrastructure, and support human-centered urban development.

The findings and insights presented in this book hold relevance for a wide audience, including scholars, researchers, practitioners, and policymakers engaged in urbanism, urban geography, urban studies, planning, innovation, and sustainability. This book serves as a reference for future research and policymaking by outlining real-world applications and their effects, promoting more investigation into the ways in which ICTs can be used to create resilient and sustainable cities.

By showing how digital innovation can boost sustainable urban transformation, this collection of case study examples ultimately aims to close the gap between theory and practice. The lessons from Hainan's ICT-based interventions offer important insights into how technology will shape the future of sustainable urban living as cities around the world attempt to manage the challenges of rapid urbanization and environmental change.

Ali Cheshmehzangi
Head of School
The University of Queensland
Brisbane, Australia

Acknowledgements This book started at one of the most challenging times: in the midst of an almost never-ending pandemic at the time (or at least, that is how it felt back then), when political tensions escalated to a new level (or else political snakes would have never learned from their actions and inactions), and where we felt safety started to have a different meaning (or else we would have stayed in the same place for another decade or so!). The so-called comfort zone started to strangulate the ones at ease, but with endless compassion, while the ones who were in unbroken agony had to pave their paths through with their bleeding fingers and flattening nails. What devastation, I recall saying it at the early stages of this journey back in April 2016. It all happened so fast, under constant pressure, with extended—yet naïve—trust between the parties, or one rambling—yet rootless—army against a solid soul of thousands years of stature, virtues, and triumph; and all in a process where things were cunningly concocted. There was never a genuine appeal plan, but all entrenched and enchanted in a crafted delusion to avoid continuous growth and success.

At the same time, scoundrels of many sorts were amalgamated with no mind, against the right flow of 'care, courage, and passion'; the care that once I said were intentionally taking out of our hearts in many ways and in multiple waves. The little boy who never succeeded in growing up knew very well how foolish and devil-alike he became just to ensure he could consciously and consistently sabotage the care that once built many opportunities for many. His lack of innovation was meant to belittle the ones who he found as obstructions to his imaginary growth; a growth that we never managed to witness, see, or perceive in any form. Alas, the care was eventually taken out of our hearts; yes, very carelessly, and indeed, very crudely, if not with vivid cruelty. But, the care and passion that are yet to be seen over subsequent deconstructions has revived healthily and has healed beyond expectations.

What an overhaul it was, of course. Yet, those who envy you attempted many times and were exasperated many times, too. Alas, they tried very hard until redemption became brisken healing, and the essential healing became a spirited recharge to extend the original care and passion once again. We had to remember that '*their aspiration often works unfaithfully into severe escalations, vending immense limitations*'.

A few months later, everything was at a halt all of a sudden, everything changed, and our gray paths became crisp and clear. We no longer had to walk on sharp edges with our bleeding bare feet. The combined tears and laughter resembled everything at a time: joy, anxiety, sorrow, mischief, vice, ignorance, reprisal, and many more. As Robert J. Oppenheimer put it well in a 1965 NBC News Documentary:

> We knew the world would not be the same. A few people laughed; a few people cried. Most people were silent. I remembered the line from the Hindu scripture, the Bhagavad Gita; Vishnu is trying to persuade the prince that he should do his duty, and to impress him, takes on his multiarmed form and says, 'Now I am become Death, the destroyer of worlds.' I suppose we all thought that, one way or another.

Here we are now! After almost two years, I'd like to warmly thank our collaborators and participants for our events (including, but not limited to, workshops, webinars, and report review sessions) in the southern regions of Hainan Island in 2021 and 2022. We first arrived there in April 2021, and we last stayed in November 2022.

The sin island of 2016 to 2019 turned out to be a place where we could cultivate and revive from time to time. We hope to continue visiting places that we aim to have a genuinely positive impact on, and not just leave them with memories of the past. You once rocked someone's world before, and you then changed everything for us in one go. We chose Hainan due to its limitations and adversities, yet we see immense opportunities for order and progress. Our multiple visits and collaborative work led to greater opportunities and connections that are still standing strong today.

Once again, only those who have endured privations are exceptionally cherished as our sincere comrades.

Competing Interests The author has no competing interests to declare that are relevant to the content of this manuscript.

About This Book

This book is a collection of context-specific ICT-based interventions to achieve sustainability from various perspectives or dimensions. The book is based on case study examples in the strategic context of Hainan, South China. Five ICT-based pilot studies were conducted, surveyed, and analysed in 2021 and 2022. The studies explore impacts on four main sustainability dimensions of environmental, social, economic, and institutional. Several ICT-based interventions are suggested to enhance environmental protection, promote health and support in elderly communities, augment social media for place promotion, create online opportunities for local markets, and help boost local tourism industries.

Urban Innovation and Sustainability is an attempt to highlight the positive side of ICT-based interventions in cities and communities. We also need to note the negative side of ICTs, which are partly covered in the case study examples. However, this book focuses on case study pilot examples to promote the nexus between innovation and sustainability. It is essential to explore opportunities that could later be scaled up, transform practices, and help develop context-specific policies. In essence, paradigm shifts, infrastructural development, and human-centric development are necessary. This book's findings interest scholars/researchers, practitioners, and authorities in various disciplines of urbanism, urban/human geography, urban studies, planning, innovation, and sustainability.

HAINAN

Contents

About the Author

Ali Cheshmehzangi is a Professor and Head of the School of Architecture, Design and Planning (ADP) at The University of Queensland, where he leads the School and researches across the fields of architecture, design, and urbanism. He has been in the World's top 2 percent field leader since 2022, recognised by Stanford University. He is among the top 25 global scholars in the urban sustainability research area.

Ali has a wealth of experience, expertise, and a strong commitment to innovation and excellence in the fields of architecture, design, and planning, the three core disciplines of his School at UQ. With a career spanning over two decades, he has made significant contributions to the academic and professional communities, with a focus on sustainable and environmentally conscious design. Prior to joining UQ, Ali held several strategic leadership and senior managerial roles, such as Vice-President for International Engagement and Global Partnership, Founding Director and Head of the Center for Innovation in Education and Research, Head of the Department of Architecture and Built Environment, Founding Director of the Urban Innovation Lab, Director of a university-wide Teaching and Learning platform, Director of International Research Network for Rural and Urban Development, Head of Research Group for Sustainable Built Environment, co-director of university-wide research priority areas, Director of Center for Sustainable Energy Technologies, and Interim Director of Digital Design Lab. So far, Ali has published over 500 journal papers, articles, conference papers, book chapters, and reports. This is his 40th academic book. Some of his previous books have received awards at the national, provincial, and municipal levels. He also has received international awards and recognition for his research on urban resilience studies and sustainability research, as well as a Vice-Chancellor's award for his impactful contribution to higher education.

Chapter 1
Digital Pathways to Sustainable Cities: An Introduction to ICT Interventions

1.1 Introduction: Digitalization and Sustainable Cities

The rapid advancement of digital technologies is transforming urban landscapes worldwide, making cities smarter, more efficient, and better equipped to handle contemporary challenges [28]. The development of smart cities—urban settings that use technology to improve living conditions, maximize resource use, and advance sustainability [29]—has been made easier by the growing reliance on digital solutions like the Internet of Things (IoT), artificial intelligence (AI), big data analytics, and blockchain technology [39].

Since more than half of the world's population lives in cities, urban areas are experiencing unprecedented growth. Cities are under tremendous pressure to preserve liveability, lessen their impact on the environment, and guarantee economic prosperity as urbanization picks up speed [49]. By facilitating data-driven decision-making, enhancing public services, and promoting connectivity, digitalization provides answers to these problems [15]. For instance, intelligent transportation systems ease traffic, smart grids maximize energy use, and digital governance enhances public participation and citizen engagement.

Furthermore, the incorporation of digital infrastructure into urban development improves predictive analysis and real-time monitoring, allowing cities to address problems before they become more serious. Technologies like edge computing and 5G networks increase communication efficiency and enable smooth connectivity between different urban sectors. IoT-enabled smart buildings improve energy management by modifying lighting, heating, and cooling systems in response to occupancy trends [38]. Additionally, the development of urban informatics, which allows cities to use massive data sets to formulate policies, aids in the optimization of urban planning and resource allocation.

Beyond just improving technology, digitalization has the potential to change governance and urban culture. Cities encourage increased cooperation between

A. Cheshmehzangi, *Urban Innovation and Sustainability*, Urban Sustainability,
https://doi.org/10.1007/978-981-96-6370-5_1

public institutions, private businesses, and citizens by implementing digital platforms. The goal of digital transformation is to transform cities into more responsive, sustainable, and inclusive ecosystems, not just more efficient ones. The rise of digital twin technology, which creates virtual models of real-world urban systems [18], is another example of how cities are adopting digitalization to model, evaluate, and improve their infrastructure.

1.2 The Role of ICT in Addressing Urban Sustainability Challenges

Beyond just improving technology, digitalization has the potential to change governance and urban culture. Cities encourage increased cooperation between public institutions, private businesses, and citizens by implementing digital platforms. The goal of digital transformation is to transform cities into more responsive, sustainable, and inclusive ecosystems, not just more efficient ones. The rise of digital twin technology, which creates virtual models of real-world urban systems, is another example of how cities are adopting digitalization to model, evaluate, and improve their infrastructure.

First, ICT promotes resource conservation, optimizes energy efficiency, and makes smart infrastructure possible, all of which contribute to environmental sustainability. For example, smart grids help control electricity use, which lowers carbon emissions overall. Additionally, ICT-driven solutions such as intelligent transport systems (ITS) and real-time air quality monitoring contribute to cleaner, more sustainable urban environments. By lowering waste and guaranteeing sustainable consumption, these technologies also help improve water management through smart metering and leak detection. Second, by making basic services like healthcare, education, and public transportation more accessible, ICT promotes social inclusivity. E-governance initiatives empower citizens by providing transparent and efficient digital platforms for communication and service delivery. Additionally, by bridging the digital divide, digital inclusion initiatives guarantee that underserved communities have access to opportunities and information.

Finally, ICT fosters economic development by enabling smart industries, digital entrepreneurship, and innovative business models. Businesses can improve operations, boost productivity, and generate new job opportunities by utilizing data analytics, automation, and artificial intelligence. Moreover, ICT-driven smart city initiatives attract investment and improve overall urban resilience [44]. In conclusion, by combining environmental, social, and economic aspects, ICT plays a crucial role in facilitating sustainability and assisting cities in their transition to a more resilient and sustainable future.

1.2.1 *Environmental Sustainability*

Better monitoring, management, and conservation of natural resources are made possible by ICT, which promotes environmental sustainability [40, 53]. By integrating renewable energy sources, lowering carbon emissions, and balancing supply and demand, smart grids and energy management systems increase energy efficiency. Similar to this, real-time sensors and predictive analytics are used by smart water management systems to find leaks, maximize distribution, and reduce waste. Waste management is also aided by digitalization, smart trash cans and tracking devices guarantee more effective recycling and collection.

Additionally, cities are using digital tools to improve climate resilience and air quality monitoring. Systems driven by AI examine trends in air pollution, assisting decision-makers in putting effective mitigation plans into action [6]. In order to improve climate resilience, urban planners create green areas and flood control systems using geographic information systems (GIS) and remote sensing technologies. The larger goal of creating more environmentally conscious and sustainable cities is furthered by these digital interventions.

1.2.2 *Social Inclusivity*

Beyond environmental issues, ICT improves accessibility, inclusivity, and quality of life, which has a revolutionary effect on social sustainability [7, 35]. Through digital platforms, e-governance initiatives enable citizens to interact with municipal authorities and promote transparent and participatory decision-making. Through telemedicine and remote diagnostics, smart healthcare systems increase access to medical services, especially in underprivileged communities. Digital education platforms also help close the knowledge gap by giving underserved groups access to online learning opportunities.

Additionally, ICT is essential to improving accessibility for individuals with disabilities in urban areas. AI-powered navigation apps and real-time public transportation tracking are examples of assistive technologies that contribute to the development of more inclusive urban environments [12]. Public safety is also improved by digital solutions [10, 24, 30], predictive policing, emergency response apps, and smart surveillance systems all increase urban security.

1.2.3 *Economic Development*

One of the main forces behind sustainable urban growth is the digital economy. By boosting productivity, encouraging entrepreneurship, and opening up new job opportunities, ICT promotes economic resilience. Businesses, particularly small and medium-sized enterprises (SMEs), can now reach a wider audience thanks to the growth of digital platforms. Financial inclusion is made easier by smart financial services like blockchain-based transactions and mobile banking, especially in emerging cities [42]. Additionally, ride-sharing, driverless cars, and intelligent public transportation are examples of ICT-driven mobility innovations that lower travel expenses, lessen carbon emissions, and enhance urban mobility [19]. Through predictive analytics and real-time tracking, digitalization also improves supply chain efficiency, supporting sustainable patterns of production and consumption.

1.2.4 Institutional Growth

Cities' institutional frameworks and governance structures are changing as a result of digitalization. ICT integration in urban administration improves decision-making processes' efficiency, accountability, and transparency. Municipalities can increase responsiveness to citizen needs, automate bureaucratic processes, and streamline services with the help of digital governance platforms. Evidence-based policymaking is further supported by open data initiatives and AI-driven analytics, which assist governments in putting specific urban sustainability plans into action [13]. Additionally, cross-sectoral partnerships and inter-agency collaboration are fostered by digital transformation. Cloud-based systems make it easy for various government agencies to share data, which improves coordination in disaster response, infrastructure management, and urban planning. By facilitating automated service delivery, predictive maintenance, and smart budgeting, digitalization also increases institutional capacity and guarantees that cities run effectively and fairly.

Cities must also address issues with data security, privacy, and the ethical application of AI as digital tools become more and more integrated into urban governance [43]. Strong protections are required to preserve citizens' rights while guaranteeing smooth digital operations due to the increasing reliance on smart technologies and data-driven decision-making. Urban ICT infrastructures may be exposed to cyberattacks [9], data breaches, and the misuse of personal data if strong security measures are not in place [26, 54]. This would erode public confidence in digital governance.

Maintaining institutional growth in the digital age will depend on ensuring digital trust through robust regulatory frameworks and cybersecurity measures. Clear data governance policies that prioritize accountability, transparency, and the responsible application of AI must be put in place by governments and urban planners. This entails implementing strict cybersecurity procedures, creating data-sharing agreements with ethical considerations, and embracing privacy-by-design principles [4]. Furthermore, to guarantee that smart city initiatives are in line with the rights and values of citizens, public participation in digital governance is crucial. Cities can fully utilize ICT while preserving public trust in technological advancements by creating a safe and welcoming digital environment.

1.3 Key Themes on Digital Pathways to Sustainable Cities

With an emphasis on ICT-driven interventions and an examination of how technology-driven solutions can tackle the urgent problems of urbanization, this book explores the complex relationship between digitalization and sustainable cities [25, 48]. Digitalization has become a transformative force in creating resilient and sustainable urban environments as cities around the world deal with mounting pressures from population growth, climate change, resource depletion, and socioeconomic inequality [3]. Cities can optimize resource management, improve public services, and raise people's quality of life [58] by utilizing smart technologies, data analytics, and artificial intelligence.

This book looks at how ICT solutions enable energy-efficient systems, smart infrastructure, and real-time environmental monitoring, all of which contribute to various sustainability factors, including environmental sustainability. By increasing access to necessary services, enabling digital governance, and encouraging public participation in urban decision-making, it also emphasizes how digital tools improve social inclusivity. The book also examines the economic effects of digital transformation, highlighting the ways in which ICT stimulates economic resilience, innovation, and new business opportunities. It also addresses the continuous institutional development that forms the fundamental basis for strengthening cities and smart urban systems via improved procedures and astute governance.

The use of smart technologies in infrastructure management and urban planning, digital governance and citizen involvement in sustainable city development, ethical AI, data security, and privacy issues in urban digitalization, and the influence of ICT on economic resilience and innovation in smart cities are some of the global trends. However, some important themes are recommended for additional research, enhancing the connection between sustainability and innovation. The following is a brief discussion of these specific themes (Fig. 1.1).

1.3.1 Smart Governance and Digital Participation

In order to shape the future of sustainable and smart cities, this subject area investigates how digital tools improve governance, urban planning, and citizen engagement. Digital governance solutions are essential for increasing administrative efficiency, promoting more inclusive decision-making, and enhancing transparency in increasingly complex urban areas. Real-time communication between citizens and authorities is made possible by e-governance platforms, which promotes greater accountability in urban management, better responsiveness to public concerns, and more effective service delivery [33, 55]. The subject also explores the idea of "digital democracy," which holds that citizens can actively engage in governance through digital voting systems, participatory budgeting, and online consultations. Cities can

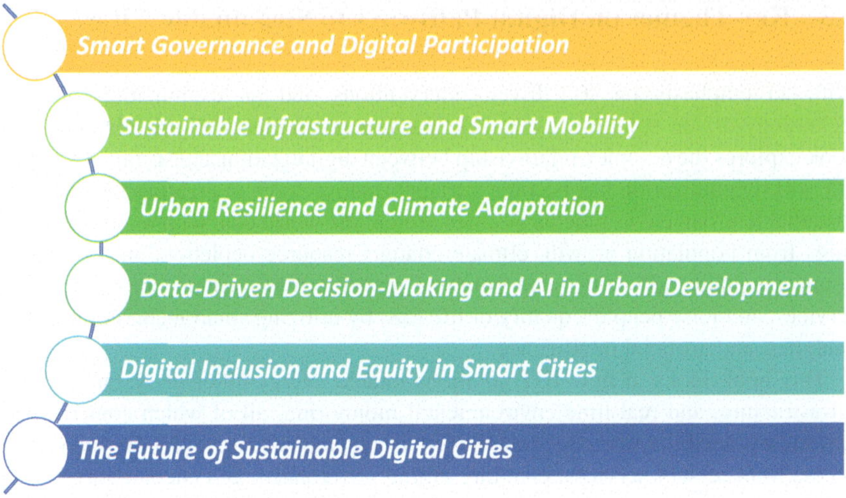

Fig. 1.1 Key themes on digital pathways to sustainable cities

make sure that policy choices and urban planning take into account the interests and needs of various communities by utilizing these tools.

Additionally, the function of artificial intelligence (AI) in policymaking is examined, emphasizing the ways in which AI-powered data analytics can forecast urban trends, optimize resource allocation, and assist evidence-based decision-making [8]. The ethical implications of AI in governance are also covered in this chapter, with a focus on the significance of openness, equity, and privacy preservation. Analyzed are best practices from the world's top smart cities, which highlight effective applications of digital governance models, lessons discovered, and tactics for expanding ICT-driven urban management solutions to improve resilience and sustainability.

1.3.2 Sustainable Infrastructure and Smart Mobility

Urban sustainability is largely dependent on mobility and infrastructure, which have an impact on environmental performance, economic productivity, and general quality of life [2]. Cities face increasing problems as they expand, including aging infrastructure, carbon emissions, and traffic congestion [50]. In order to help cities operate more effectively and sustainably, this section looks at how ICT can be used to optimize transportation networks, create smart grids, and guarantee resilient urban infrastructure. For instance, IoT sensors, real-time data, and AI-powered analytics are all used by smart transportation systems to increase mobility, lessen traffic, and boost the effectiveness of public transportation. Connected car technology and adaptive traffic signal control are two examples of intelligent transportation solutions that help reduce emissions and improve traffic flow. Furthermore, ICT promotes the growth of electric mobility by making it easier to build infrastructure for EV charging and incorporate renewable energy sources into transportation networks.

The use of digital twins, or virtual copies of physical infrastructure, in resilience-building and urban planning is also examined in this field. City planners can test various scenarios, anticipate infrastructure vulnerabilities, and optimize resource allocation with the help of these sophisticated simulation tools. Cities can design more flexible, effective, and sustainable urban environments by incorporating ICT into mobility and infrastructure planning. Insights into the future of urban development and sustainability can be gained from case studies on digital twins, electric mobility, and smart transportation systems.

1.3.3 Urban Resilience and Climate Adaptation

With cities facing a serious threat from climate change, digital solutions are becoming a crucial part of climate adaptation plans, making cities more resilient to resource scarcity, extreme weather, and rising temperatures [45]. Cities must incorporate smart technologies into their sustainability frameworks because the risks associated with climate change have increased due to rapid urbanization and environmental degradation. This theme area explores the applications of IoT, AI, and big data in disaster relief, climate monitoring, and green infrastructure planning, highlighting creative strategies that improve urban resilience.

Through sensor networks that monitor water levels, temperature variations, and air quality, IoT makes it possible to monitor the environment in real time. Cities can take proactive mitigation measures by using these data-driven insights to predict climate-related issues like pollution, heat waves, and flooding. By analyzing enormous datasets to model climate trends, evaluate vulnerabilities, and suggest adaptive measures, AI-powered predictive analytics further improve these capabilities.

Digital solutions are essential for emergency response coordination and early warning systems in disaster management (Prguba and Garaplija, 2023). While geospatial mapping tools aid in the efficient planning of evacuation routes and resource allocation, AI-driven simulations aid in the prediction of natural disasters. As urban planners use big data to create climate-resilient spaces like permeable pavements, urban forests, and green roofs, digital innovations also help with green infrastructure planning. City officials can ensure the most sustainable and effective solutions by simulating the effects of climate interventions using digital twin technology. Cities can improve their climate adaptation plans and create a more sustainable future by utilizing ICT.

1.3.4 Data-Driven Decision-Making and AI in Urban Development

Data analytics and artificial intelligence (AI) are essential for improving urban services, changing how cities function, and adapting to changing issues. AI-driven solutions facilitate more effective resource management, improve service delivery, and support evidence-based decision-making as urban populations increase and sustainability concerns grow. This theme looks at machine learning applications, predictive modelling, and how data governance guarantees moral and effective urban management.

City planners can foresee future trends like population growth, transportation demand, and energy consumption by using predictive modelling. AI-driven models can optimize urban services, lowering inefficiencies and enhancing public infrastructure by examining historical data. For example, real-time traffic signal adjustments made by predictive analytics in traffic management can help avoid congestion, and AI-powered energy grids can balance supply and demand for electricity, cutting down on waste and fostering sustainability. By facilitating automation and adaptive decision-making, machine learning applications further improve city services. For instance, smart sensors are used by AI-driven waste management systems to optimize collection routes, lowering expenses and their negative effects on the environment. Predictive AI technologies in healthcare can assist in detecting disease outbreaks, enabling prompt action and better public health results.

However, there are serious concerns regarding data governance, privacy, and ethical issues brought up by the growing reliance on AI and data analytics. To avoid biases and guarantee fairness in decision-making, cities must set up strong frameworks for data transparency, security, and responsible AI use. Cities can fully utilize AI while upholding public confidence and encouraging sustainable development by fusing data-driven urban planning with ethical AI practices.

1.3.5 Digital Inclusion and Equity in Smart Cities

Although there are many advantages to digitalization in terms of sustainability, convenience, and efficiency, there is also a chance that social inequality will worsen. The digital divide—unequal access to digital tools and resources—becomes a critical issue as cities adopt smart technologies, potentially further marginalizing vulnerable populations. This theme area covers methods for guaranteeing digital inclusion, closing the digital divide, and developing fair smart city solutions that serve all residents, irrespective of their socioeconomic background, age, gender, or location.

Access to reasonably priced internet services must be given top priority by cities in order to guarantee digital inclusion, particularly in underprivileged areas. Providing low-cost or subsidized digital devices and expanding broadband infrastructure are essential first steps in bringing underserved communities online. In order to equip people with the knowledge and abilities necessary to use online platforms, access e-governance services, and engage in the digital economy, local governments and organizations must also fund digital literacy initiatives.

Addressing the particular needs of vulnerable populations, such as the elderly, low-income families, and people with disabilities, is another aspect of closing the digital divide. Cities should take these needs into account when creating inclusive smart city solutions, providing digital services, accessible websites, and assistive technologies that are suited to a range of user needs. Furthermore, to guarantee that digital tools represent the priorities of all communities, it is crucial to promote citizen participation in the creation of smart city initiatives. Co-design procedures, participatory governance models, and public consultations can all contribute to the development of more responsive, equitable, and inclusive urban systems. By implementing these tactics, cities can use digital transformation to advance social justice and guarantee that every citizen takes advantage of the opportunities presented by a connected, digital world.

1.3.6 The Future of Sustainable Digital Cities

The last subject area covered here looks to the future, taking into account the new developments in urban digitalization that have the potential to change how cities operate, interact with their residents, and handle sustainability issues. New opportunities and challenges will arise as digital technologies develop further, necessitating creative approaches to urban governance and planning. The function of the metaverse in urban planning, the moral ramifications of AI-powered cities, and the influence of new technologies on urban sustainability are some of the subjects covered in this particular theme area.

Urban planning is starting to be influenced by the idea of the metaverse, a virtual reality environment where the real and virtual worlds collide. Future urban development projects could be simulated by cities using the metaverse, giving citizens and planners a chance to envision changes before they are put into action. Because stakeholders can examine suggested urban changes—from public areas to infrastructure—in a virtual setting, this immersive experience can promote greater community engagement. Additionally, the metaverse might be used as a test site for cutting-edge urban ideas like sustainable transportation and intelligent zoning.

The ethical ramifications of AI-driven cities are also covered in the course, as these systems may have an impact on choices about resource allocation, law enforcement, and urban administration. Even though AI could increase productivity, privacy, bias, and accountability are issues. The emergence of autonomous decision-making in urban areas necessitates a thorough review of ethical standards to guarantee that AI is applied in a way that respects and benefits all citizens.

Lastly, the effect of new technologies on urban sustainability is examined, including blockchain, 5G, and IoT. These technologies help cities adapt to climate change and advance long-term sustainability by providing game-changing solutions for waste reduction, energy optimization, and sustainable mobility. The future of urban digitalization will be shaped by these new trends taken together, which present both opportunities and difficulties for sustainable, just cities.

The relationship between technology and sustainability is becoming more and more important as cities all over the world embrace digital transformation. Strong tools for tackling urban issues, increasing productivity, and promoting inclusivity are provided by digitalization. But it also calls for meticulous preparation, moral reflection, and a dedication to digital equity. In order to ensure that cities are resilient, creative, and human-centered, this book attempts to present a thorough analysis of the ways in which digital solutions can support sustainable urban futures. This work aims to assist researchers, policymakers, and urban planners in utilizing digitalization for sustainable urban development by examining empirical examples, theoretical insights, and policy recommendations.

1.4 The Evolution of Smart and Sustainable Cities

The history of urban planning, technological development, and sustainability requirements are all intricately entwined with the development of smart and sustainable cities. In order to accommodate expanding populations and economic activity, urban planning has historically focused on the concepts of spatial organization, infrastructure development, and resource allocation. Urban landscapes underwent a dramatic change during the Industrial Revolution as a result of mass production, increased migration to city centers, and rapid urbanization. A change to contemporary urban planning paradigms that integrated zoning regulations, green spaces, and transportation networks was necessary, though, as these developments also brought about traffic jams, pollution, and public health emergencies. Cities started incorporating systemic planning techniques that prioritized efficiency, connectivity, and quality of life by the middle of the twentieth century, but many of these tactics stayed stagnant and were reactive rather than proactive in tackling new urban issues.

Cities entered a new stage of development marked by the incorporation of information and communication technologies (ICT) into urban systems as a result of the introduction of digital technologies in the late 20th and early twenty-first centuries. This shift paved the way for the rise of smart cities, which improve urban management and service delivery by utilizing data analytics, artificial intelligence, and networked infrastructure. As governments and urban planners realized how digital solutions could optimize energy consumption, enhance public transportation, and promote participatory governance, the idea of "smart cities" gained traction. Crucially, sustainability goals are becoming more and more entwined with smart city initiatives, bringing technological advancements into line with social and environmental demands. According to international frameworks like the Sustainable Development Goals (SDGs) of the UN, sustainable urban development places a strong emphasis on the need to lower carbon footprints, increase climate change resilience, and foster inclusive economic growth. Municipalities can reduce ecological degradation, promote economic competitiveness, and improve the general well-being of their citizens by integrating sustainability into smart city initiatives.

In order to solve urgent urban issues, ICT-based solutions were frequently implemented in early smart city development interventions. The deployment of intelligent transport systems (ITS) in places like Singapore and London is among the most prominent examples. In order to control traffic congestion and encourage effective mobility, Singapore, for example, developed a dynamic road pricing system in the 1990s using real-time data. Similar to this, London's 2003 implementation of a congestion charge system used ICT to monitor and control traffic flow, resulting in fewer emissions and traffic jams. The Songdo International Business District in South Korea, which was intended to be a fully integrated smart city with embedded digital infrastructure, is another noteworthy case study. Songdo, a model for future urban developments, was created with sustainability in mind and integrates green building technologies, smart grids, and waste management automation. Barcelona has also led the way in smart city innovation, using sensor-based technologies to maximize waste collection, energy efficiency, and water management [51, 47]. By lowering environmental impacts and enhancing resource use, these pioneering projects show how digital transformation has strengthened sustainability objectives [56] while also increasing urban efficiency.

The combination of sustainability and smart technologies will be essential to solving today's urban problems as cities continue to change. Predictive urban management is made possible by the combination of artificial intelligence, the Internet of Things (IoT), and big data analytics. This promotes resilience against socioeconomic inequality, resource scarcity, and climate change. To guarantee inclusive and equitable results, the shift to smart and sustainable cities calls for cautious governance, policy alignment, and community involvement. Policymakers and urban planners can create flexible, forward-thinking plans that capitalize on technological advancements while preserving ecological and social integrity by learning from past urban transitions and early ICT interventions. Ultimately, the development of smart and sustainable cities is a reflection of a continuous shift toward more intelligent, effective, and resilient urban settings, supported by the dynamic interaction between sustainability and digitalization.

1.4.1 Key ICT Interventions for Urban Sustainability

The way cities are planned, run, and observed has changed dramatically as a result of the incorporation of information and communication technologies (ICT) into urban sustainability initiatives. The optimization of urban infrastructure and services is made possible by the more accurate, real-time data that emerging digital tools and frameworks are giving decision-makers. Digital twin technology, AI-powered urban management systems, and GIS-enabled spatial analytics are some of the most revolutionary ICT interventions. By increasing productivity, lessening their negative effects on the environment, and fostering resilience in urban environments across the globe, these technologies are essential to improving sustainability.

Digital twin technology, which provides a virtual representation of physical environments that can simulate and predict urban dynamics, has become a potent tool for real-time urban monitoring. In order to create a dynamic visualization of cities and enable better decision-making in urban development, these digital replicas combine enormous volumes of data from sensors, satellites, and historical documents. The Virtual Singapore project in Singapore is a prime example of the potential of digital twins since it provides a high-fidelity, data-driven model of the city that allows officials to evaluate environmental conditions, optimize energy use, and analyze urban mobility. Similarly, by simulating various scenarios and assessing possible interventions, the Rotterdam digital twin in Europe helps with flood risk assessment and climate adaptation planning. By facilitating quick reactions to environmental threats, maximizing resource allocation, and bolstering data-driven governance, such real-time monitoring capabilities improve urban resilience.

By facilitating automated decision-making and predictive analytics, artificial intelligence (AI), big data, and the Internet of Things (IoT) have completely transformed urban planning and management. Large volumes of urban data can be gathered, processed, and analyzed thanks to these technologies, which makes waste management, energy distribution, and transportation systems more effective. Real-time data from IoT-enabled cameras and sensors is used by AI-powered traffic management systems in Chinese cities like Hangzhou to dynamically control traffic flows, lowering emissions and congestion. By optimizing energy consumption based on real-time supply and demand patterns, big data analytics in Copenhagen aid in the development of smart grids. As demonstrated in Barcelona, where smart waste bins with sensors detect fill levels and optimize collection routes, lowering fuel consumption and operating costs, the application of AI and IoT in waste management has also proven successful. By lowering carbon footprints, minimizing resource waste, and improving the effectiveness of urban services, these interventions support sustainability.

Spatial analytics and Geographic Information Systems (GIS) are essential tools for sustainable urban design because they enable data-driven planning procedures and spatially-informed decision-making. Urban planners can evaluate environmental risks, examine land-use patterns, and create resilient infrastructure with the help of GIS technology. To ensure climate resilience, New York City, for example, uses GIS-based flood mapping to inform zoning laws and infrastructure investments in flood-prone areas. In order to improve biodiversity and lessen the impact of the urban heat island, parks, green roofs, and urban forests are optimally integrated into green infrastructure networks in Melbourne using GIS tools. Additionally, by mapping service gaps and organizing infrastructure interventions that support equitable urban development, spatial analytics help Nairobi's informal settlement upgrading initiatives. Cities can create more adaptable and sustainable urban environments that meet social demands and ecological constraints by incorporating GIS into planning procedures.

The intersection of GIS-enabled spatial analytics, AI-driven urban management, and digital twin technology highlights how ICT can revolutionize urban sustainability. Cities can now better monitor environmental conditions, use resources efficiently, and create resilient infrastructure thanks to these technologies. Policymakers, researchers, and tech developers must work together to make sure that ICT interventions are inclusive, scalable, and accessible for these tools to be successfully incorporated into urban governance frameworks. Using digital solutions will be crucial to creating resilient and sustainable urban futures as cities continue to struggle with issues like population growth, resource scarcity, and climate change. Cities can increase productivity, reduce their environmental impact, and promote sustainable urban development globally by adopting ICT-driven innovations.

1.5 Governance and Policy in the Digital Urban Era

Cities are being developed and managed differently as a result of the new paradigms in governance and policy brought about by the digital urban era. Sustainable urban environments are greatly aided by digital governance, which is defined by data-driven decision-making, networked technologies, and participatory mechanisms. In order to guarantee inclusivity, transparency, and moral accountability, governance models must change as cities incorporate smart technologies and artificial intelligence more and more into their administrative frameworks. Opportunities and difficulties arise when technology, policy, and governance come together in urban settings, especially when it comes to open data projects, sustainability efforts, and the moral implications of smart urbanism.

By utilizing technological innovations that facilitate real-time monitoring, predictive analytics, and data-driven planning, digital governance improves sustainable decision-making. Urban administrations can improve resilience against environmental and socioeconomic challenges, minimize inefficiencies, and optimize resource allocation by leveraging big data, artificial intelligence, and the Internet of Things (IoT). For example, by dynamically adjusting supply and demand, smart grids and intelligent transportation systems help to reduce carbon emissions and increase energy efficiency. Furthermore, by encouraging cooperation among diverse stakeholders, digital platforms support integrated urban planning and guarantee that infrastructure development is in line with sustainability objectives. Through accurate data analytics, policymakers can manage land-use patterns, evaluate environmental impacts, and improve disaster preparedness by utilizing Geographic Information Systems (GIS). More informed decision-making that puts long-term ecological balance, economic viability, and social well-being first is made possible by the capacity to synthesize diverse datasets into actionable insights.

In the digital age, open data, transparency, and citizen participation are essential for promoting democratic and accountable urban governance. Data accessibility enables businesses, researchers, and citizens to actively participate in urban decision-making processes. Governments can promote evidence-based policymaking, foster innovation, and increase public-authority trust by making datasets publicly accessible. Digital platforms that offer real-time updates on municipal projects, budgeting, and environmental indicators serve to further strengthen transparency in governance. Citizens can now directly contribute their opinions and insights to policy discussions thanks to the rise of participatory governance models like crowdsourcing projects and e-governance portals. Digital tools, such as online consultations and mobile applications, enable ongoing civic engagement and the co-creation of solutions that tackle issues unique to a community. However, the quality, usability, and accessibility of the information provided determine how successful open data initiatives are. For policymakers, ensuring that data is integrated into decision-making processes, protected from misuse, and presented in understandable formats continues to be a crucial challenge.

The swift development of smart urbanism brings with it ethical and policy issues that need to be carefully navigated. Concerns about privacy, security, and digital rights are raised by the use of surveillance technologies, biometric data collection, and algorithm-driven decision-making. If algorithmic biases are not addressed by ethical frameworks and regulatory oversight, the growing reliance on artificial intelligence in urban management could reinforce social inequality. Additionally, concerns regarding data ownership, governance, and the impact of profit-driven incentives on public services are brought up by technology companies' privatization of urban digital infrastructure. A persistent policy challenge is ensuring fair access to digital tools and avoiding digital exclusion, especially for underserved communities. Furthermore, strict regulatory measures are required to prevent the misuse of personal information due to the ethical implications of data collection and utilization. In the digital urban age, preserving public confidence and respecting democratic values require the establishment of unambiguous rules regarding data governance, accountability, and moral AI application.

In the age of digital cities, governance, and policy must change to take advantage of technology's advantages while tackling its drawbacks. Building resilient and inclusive cities requires a well-rounded strategy that incorporates sustainable decision-making, encourages openness and participation, and respects moral principles. Technological developments require proactive policymakers to maintain regulatory frameworks' flexibility in response to new trends while defending fundamental rights. As long as it is supported by strong governance frameworks and moral principles, the digital transformation of urban governance offers a previously unheard-of chance to improve effectiveness, sustainability, and citizen empowerment. The focus must continue to be on utilizing digital innovations to promote a more sustainable and equitable future as cities continue to adopt smart urbanism.

1.6 Barriers and Challenges in Digital Sustainability

In order to promote long-term social, economic, and environmental well-being, digital sustainability as a concept includes the ethical and responsible development of digital technologies [41]. However, there are many obstacles and difficulties in achieving digital sustainability, which need to be carefully considered. These difficulties include the complicated integration of digital and physical systems, cybersecurity and data privacy issues in smart city infrastructures, and the digital divide and urban inequality [11, 27, 32]. To solve these problems and make sure that digital sustainability is in line with more general sustainable development objectives, a multi-pronged strategy that takes into account technological developments [37], policy interventions [34], and inclusive governance frameworks is needed.

The ongoing digital divide, which exacerbates already-existing urban inequalities, is one of the main obstacles to digital sustainability [20–22, 31, 46]. Disparities in access to digital resources pose serious obstacles for marginalized communities, despite the fact that digital transformation is frequently hailed as a way to foster social inclusion and economic growth. High-speed internet, cutting-edge technology infrastructure, and digital literacy initiatives benefit wealthier populations in many urban settings, while underprivileged groups experience restricted access, financial limitations, and a lack of digital literacy. This disparity is especially noticeable in smart city projects, where wealthy urban areas are disproportionately served by digital services and data-driven decision-making, leaving less affluent neighbourhoods neglected. The digital divide reinforces spatial disparities within cities and goes beyond connectivity to include unequal access to digital services, e-governance platforms, and smart infrastructure. Furthermore, the expansion of digital services without all-encompassing inclusion plans runs the risk of widening social divides since people who cannot use digital systems encounter more obstacles to civic engagement, education, and economic participation. Targeted policies that place a high priority on digital literacy initiatives, reasonably priced internet access, and equitable urban planning that incorporates digital equity as a fundamental principle are required to address these disparities.

In addition to accessibility issues, cybersecurity flaws and data privacy hazards present a challenge to digital sustainability in smart city infrastructures. The swift implementation of interconnected digital systems, such as cloud computing, artificial intelligence, and sensor networks, has revolutionized urban management but also brought about serious cybersecurity risks. In order to maximize services, increase resource efficiency, and improve citizen engagement, smart city infrastructures rely on enormous volumes of data. However, cities are more vulnerable to data breaches, cyberattacks, and illegal surveillance as a result of their growing reliance on digital technologies. In addition to endangering personal privacy, cybersecurity threats also pose systemic risks to vital urban infrastructure, including public safety systems, transportation networks, and energy grids. Since smart cities are interconnected, a single security breach can have a domino effect, disrupting vital services and eroding public confidence in digital governance. Furthermore, since private technology companies and local governments frequently find it difficult to put in place strong data protection frameworks, data privacy governance continues to be a controversial topic. The landscape is further complicated by issues with data ownership, consent procedures, and ethical data usage, necessitating extensive regulatory measures to guarantee that individual rights are not compromised in the name of digital sustainability. In order to create cybersecurity resilience, a mix of technological advancements, legal frameworks, and public education campaigns are needed to enable citizens to recognize and defend their digital rights while maintaining the security and reliability of smart city infrastructures.

The difficulty of integrating digital and physical systems in urban settings is another significant obstacle to digital sustainability [1]. The shift to digitally enhanced urban systems requires smooth communication between digital technologies and the physical infrastructure that is already in place, but this integration is frequently complicated by organizational, technical, and policy issues. Many cities' legacy infrastructure was not built to support cutting-edge digital systems, which resulted in inefficiencies, compatibility problems, and expensive retrofitting procedures [52]. Furthermore, the merging of digital and physical systems necessitates concerted efforts from a variety of stakeholders, such as community organizations, technology suppliers, and local government agencies. This can result in complicated governance and competing priorities. In order to make sure that digital interventions improve rather than disturb already-existing urban ecosystems, smart technology deployment in urban areas requires careful calibration [23]. Ineffective digitalization initiatives can lead to disjointed urban systems, where discrete technological fixes fall short of producing significant gains in urban sustainability [5]. Additionally, it is becoming more widely acknowledged that digital interventions need to be context-specific because solutions that are effective in technologically advanced cities might not be practical in cities with different sociopolitical landscapes or resource constraints [14]. Strategic planning, cross-sector cooperation, and flexible urban policies that take into consideration both technological developments and the lived realities of urban populations are necessary for the successful integration of digital and physical systems.

A comprehensive strategy that tackles the interrelated problems of the digital divide, cybersecurity vulnerabilities, and system integration complexities is required to achieve digital sustainability. Targeted investments in digital literacy, reasonably priced connectivity, and inclusive smart city planning that gives priority to underserved communities are all necessary to close the digital divide. Strong legislative frameworks, technological protections, and public awareness campaigns that uphold the integrity of urban digital ecosystems while defending individual rights are necessary to ensure cybersecurity resilience and data privacy in smart city infrastructures. In the meantime, resolving issues with digital and physical system integration calls for multi-stakeholder cooperation, strategic urban planning, and context-sensitive implementation techniques that match new technology with the objectives of sustainable urban development. Digital sustainability must be pursued in a fair, safe, and integrated way that respects the values of social justice, resilience, and long-term sustainability as cities continue to embrace digital transformation.

1.7 Conclusions: Towards a Digital-Driven Sustainable Future

The transformative role of information and communication technologies (ICTs) in creating sustainable urban futures was examined in the previous discussion. Digital interventions have become essential facilitators of efficiency, resilience, and inclusivity as cities struggle with the urgent issues of climate change, rapid urbanization, and resource constraints. Important ICT applications have shown the ability to optimize resource use, lower carbon emissions, and improve service delivery. These include smart grids, intelligent transportation systems, urban sensing, and data-driven governance. ICTs have enabled evidence-based decision-making and predictive analytics, from the use of artificial intelligence in traffic management to the incorporation of digital twins in urban planning. This has strengthened cities' ability to adapt quickly to changing sustainability demands. Furthermore, citizen engagement has been redefined by the widespread use of cloud computing, Internet of Things (IoT) networks, and mobile applications. These developments have led to the development of participatory governance models that enable communities to co-create their urban environments. These developments highlight the need for digital transformation as a key component of sustainable urban growth.

Looking ahead, interdisciplinary research, ongoing technological advancements, and innovative policy will likely drive further evolution of the digital pathways in sustainable urbanism [16]. Urban modelling capabilities will be improved by the growing use of big data analytics and machine learning, providing a deeper understanding of socio-environmental dynamics and opening the door to more flexible planning strategies. Blockchain technology has the potential to improve accountability and transparency in urban governance, especially in sectors like supply chain management, energy trading, and land administration. Furthermore, hybrid strategies that combine ecological and technological resilience will be produced by the convergence of ICTs with nature-based solutions, such as ecosystem monitoring and climate-responsive urban design. To guarantee fair access to the advantages of technology for all socioeconomic groups and geographical locations, the digital divide must be addressed. Therefore, inclusivity must be given top priority in future digital pathways to ensure that digital solutions promote social justice and the well-being of all rather than worsening already-existing inequalities.

A truly digitally driven sustainable future requires innovative policymaking and interdisciplinary cooperation. The intricacy of urban sustainability issues necessitates a comprehensive strategy that cuts across sectoral lines and integrates knowledge from disciplines like behavioural economics, environmental science, computer engineering, public policy, and urban planning. To promote multi-stakeholder engagement and create synergies between governmental institutions, private businesses, academic institutions, and civil society organizations, effective governance mechanisms must be put in place [57, 20]. Adopting flexible regulatory frameworks that strike a balance between ethical issues like data privacy, cybersecurity, and digital rights and technological innovation is another imperative for policymakers [36]. In order to handle new risks and optimize the advantages of digitalization, regulatory frameworks must change as digital infrastructures become more integrated into urban ecosystems. Furthermore, to give local communities and urban professionals the tools they need to successfully navigate and capitalize on digital transformations, investments in capacity-building and digital literacy will be crucial. In this regard, advancing evidence-based strategies for sustainable urban futures will require a forward-looking research agenda that incorporates empirical studies, policy experimentation, and technological prototyping.

A digitally driven sustainable future necessitates both practical execution and visionary leadership. Technological advancements must complement more general sustainability objectives as cities continue to embrace digitalization, making sure that digital solutions strengthen social inclusion, economic resilience, and environmental stewardship [17]. Because sustainable urbanism is multi-scalar and interdisciplinary, digital interventions must be transformative processes that reshape the fundamentals of urban development rather than just small-scale enhancements. Through the promotion of cooperative governance, the development of innovative policies, and the advancement of technological frontiers, stakeholders from various sectors can work together to create sustainable and digitally intelligent urban environments.

References

1. Alao KT, Gilani SIUH, Sopian K, Alao TO (2024) A review on digital twin application in photovoltaic energy systems: challenges and opportunities. JMST Advances 6(3):257–282
2. Alberti M (1996) Measuring urban sustainability. Environ Impact Assess Rev 16(4–6):381–424
3. Allam Z, Cheshmehzangi A (2024) The green new deal as a pathway for climate and social justice. In: Sustainable Futures and Green New Deals. Cham: Springer Nature Switzerland, pp 105–133
4. Allam Z, Dhunny ZA (2019) On big data, artificial intelligence and smart cities. Cities 89:80–91
5. Allam Z, Sharma A, Cheshmehzangi A (2024) LiFE in the city: behavioural changes can drive urban sustainability goals. Cities 151:105163
6. Amiri Z, Heidari A, Navimipour N J (2024) Comprehensive survey of artificial intelligence techniques and strategies for climate change mitigation. Energy, 132827
7. Appio FP, Lima M, Paroutis S (2019) Understanding Smart Cities: Innovation ecosystems, technological advancements, and societal challenges. Technol Forecast Soc Chang 142:1–14
8. Attah RU, Garba BMP, Gil-Ozoudeh I, Iwuanyanwu O (2024) Leveraging geographic information systems and data analytics for enhanced public sector decision-making and urban planning. Magna Sci Adv Res Rev 12(2):152–163

9. Baig ZA, Szewczyk P, Valli C, Rabadia P, Hannay P, Chernyshev M, Johnstone M, Kerai P, Ibrahim A, Sansurooah K, Syed N (2017) Future challenges for smart cities: Cyber-security and digital forensics. Digit Investig 22:3–13
10. Baldini G, Karanasios S, Allen D, Vergari F (2013) Survey of wireless communication technologies for public safety. IEEE Communications Surveys & Tutorials 16(2):619–641
11. Balfaqih M, Alharbi SA (2022) Associated information and communication technologies challenges of smart city development. Sustainability 14(23):16240
12. Bayashot ZA (2024) The contribution of AI-powered mobile apps to smart city ecosystems. J Softw Eng Appl 17(3):143–154
13. Caprotti F, Cowley R, Bailey I, Joss S, Sengers F, Raven R, Spaeth P, Jolivet E, Tan-Mullins M, Cheshmehzangi A, Xie L (2017) Smart eco-city development in Europe and China: policy directions. Exeter: University of Exeter, Available from: https://eprints.gla.ac.uk/182634/
14. Cheshmehzangi A (2016) City enhancement beyond the notion of "Sustainable City": Introduction to integrated assessment for city enhancement (iACE) toolkit. Energy Procedia 104:153–158
15. Cheshmehzangi A (2022) The impacts of ICT platforms on public place use in China. Built Environment 48(2):264–279
16. Cheshmehzangi A (2022b) The role of ICT and smartness for peace-building processes and practices. In ICT, cities, and reaching positive peace. Singapore: Springer Nature Singapore, pp 17–46
17. Cheshmehzangi A (2022) ICT, cities, and reaching positive peace. Springer, Singapore
18. Cheshmehzangi A, Pourroostaei Ardakani S (2024) Navigating the Future with the Nexus Between Digital Twin Computing and Urban Intelligence. In: Pourroostaei Ardakani S, Cheshmehzangi A (eds). Digital twin computing for urban intelligence. Singapore: Springer Nature Singapore, pp 245–256
19. Cheshmehzangi A, Zou T (2024) Literary analysis of the connection between urban innovations and Sustainable Development Goals (SDGs). Sustainability 16(15):6405
20. Cheshmehzangi A, Zou T, Su Z (2022) The digital divide impacts on mental health during the COVID-19 pandemic. Brain Behav Immun 101:211–213
21. Cheshmehzangi A, Zou T, Su Z, Tang T (2023) The growing digital divide in education among primary and secondary children during the COVID-19 pandemic: An overview of social exclusion and education equality issues. J Hum Behav Soc Environ 33(3):434–449
22. Cheshmehzangi A, Zou T, Su Z, Dawodu A, Ardakani SP, Sedrez M, Tang T (2024) Growing digital divide for expatriates population in China: a user perspective analysis during the COVID-19 Pandemic. SAGE Open 14(4):21582440241274612
23. Faroqi H, Irajifar L, Cheshmehzangi A (2022) Sustainable development in smart and resilient local government areas: an empirical investigation of Victoria, Australia. Int J Sustain Dev & Plan 17(6)
24. Fraga-Lamas P, Fernández-Caramés TM, Suárez-Albela M, Castedo L, González-López M (2016) A review on internet of things for defense and public safety. Sensors 16(10):1644
25. Guma PK (2021) Rethinking Smart Urbanism: city-making and the spread of digital infrastructures in Nairobi. Eburon Uitgeverij BV
26. Habibzadeh H, Nussbaum BH, Anjomshoa F, Kantarci B, Soyata T (2019) A survey on cybersecurity, data privacy, and policy issues in cyber-physical system deployments in smart cities. Sustain Cities Soc 50:101660
27. Ismagilova E, Hughes L, Rana NP, Dwivedi, YK (2022) Security, privacy and risks within smart cities: Literature review and development of a smart city interaction framework. Inf Syst Front, 1–22
28. Kourtit K, Nijkamp P, Steenbruggen J (2017) The significance of digital data systems for smart city policy. Socioccon Plann Sci 58:13–21
29. Lea R (2017) Smart cities: An overview of the technology trends driving smart cities. IEEE Advancing technology for Humanity 3(March):1–16
30. Marusin A, Marusin A, Ablyazov T (2019, September) Transport infrastructure safety improvement based on digital technology implementation. In: International Conference on Digital

Technologies in Logistics and Infrastructure (ICDTLI 2019). Dordrecht: Atlantis Press, pp 348–352
31. Mathrani A, Umer R, Sarvesh T, Adhikari J (2023) Rural-urban, gender, and digital divides during the COVID-19 lockdown: a multi-layered study. Societies 13(5):122
32. Mazzetto S (2024) A review of urban digital twins integration, challenges, and future directions in smart city development. Sustainability 16(19):8337
33. Milakovich ME (2012) Digital governance: New technologies for improving public service and participation. Routledge, New York and London
34. Millard J (2023) Impact of digital transformation on public governance (No. JRC133975). Joint Research Centre (Seville site)
35. Morrar R, Arman H, Mousa S (2017) The fourth industrial revolution (Industry 4.0): a social innovation perspective. Technol Innov Manag Rev 7(11):12–20
36. Nguyen MT, Tran MQ (2023) Balancing security and privacy in the digital age: an in-depth analysis of legal and regulatory frameworks impacting cybersecurity practices. International Journal of Intelligent Automation and Computing 6(5):1–12
37. Omweri FS (2024) A systematic literature review of e-government implementation in developing countries: examining urban-rural disparities, institutional capacity, and socio-cultural factors in the context of local governance and progress towards SDG 16.6. Int J Res Innov Soc Sci 8(8):1173–1199
38. Pourroostaei Ardakani S, Cheshmehzangi A (2023) Big data analytics and the future of smart transport and healthcare systems. In: Big data analytics for smart transport and healthcare systems. Singapore: Springer Nature Singapore, pp 175–184
39. Pourroostaei Ardakani S, Cheshmehzangi A (eds) (2024) Digital twin computing for urban intelligence. Singapore: Springer Nature
40. Qamruzzaman M, Kor S (2024) Navigating the path to environmental sustainability: Insights from CIVETS on the intersection of ICT diffusion, natural resources, and green technological innovation. PLoS ONE 19(12):e0309264
41. Raihan A (2024) A review of the potential opportunities and challenges of the digital economy for sustainability. Innovation and Green Development 3(4):100174
42. Raslan A, Kapogiannis G, Cheshmehzangi A, Tizani W, Towey D (2020, July) Blockchain: Future facilitator of asset information modelling and management?. In: 2020 IEEE 44th Annual Computers, Software, and Applications Conference (COMPSAC). IEEE, pp 523–528
43. Rasoulzadeh S, Bababei B, Ghasemzadeh B, Irani M, Huovila A (2025) Social smart city research: interconnections between participatory governance, data privacy, artificial intelligence and ethical sustainable development. Frontiers in Sustainable Cities 6:1514040
44. Raven RPJM, Cheshmehzangi A, Jong M, Sengers F, Späth, P, Xie, L (2017) An institutional perspective on smart city experimentation:: Comparing institutional arrangements in Amsterdam, Hamburg, Ningbo and Songdo. In: The 8th International Sustainability Transitions Conference
45. Rezvani SM, de Almeida NM, Falcão MJ (2023) Climate adaptation measures for enhancing urban resilience. Buildings 13(9):2163
46. Ruiu ML, Ragnedda M (2024) Digital-environmental poverty: digital and environmental inequalities in the post-covid era. Singapore: Springer Nature
47. Salman MY, Hasar H (2023) Review on environmental aspects in smart city concept: Water, waste, air pollution and transportation smart applications using IoT techniques. Sustain Cities Soc 94:104567
48. Sibiya B (2023) Digital transformation of cities through Emerging Industry 4.0 Smart Technologies and Infrastructure in South Africa. University of Johannesburg (South Africa)
49. Sukanya R, Tantia V (2023) Urbanization and the impact on economic development. In: New perspectives and possibilities in strategic management in the 21st century: between tradition and modernity. IGI Global, pp 369–408
50. Sultana S, Salon D, Kuby M (2019) Transportation sustainability in the urban context: A comprehensive review. Urban Geogr 40(3):279–308

51. Talamo CINZIA, Pinto MR, Viola S, Atta N (2019, July) Smart cities and enabling technologies: influences on urban facility management services. In: IOP Conference Series: Earth and Environmental Science, Vol 296, No 1. IOP Publishing, p 012047
52. Tasselli N (2023) BIM digital integration for existing buildings Processes and tools to foster technological innovation of added-value building systems applied to the 20th century built asset in Emilia-Romagna. Available online from: https://sfera.unife.it/handle/11392/2515193
53. Teckshawer T (2023) The potential of information and communication technology to promote sustainable natural resources management and its unintended consequences. Technium Soc. Sci. J. 44:673
54. Telo J (2023) Smart city security threats and countermeasures in the context of emerging technologies. International Journal of Intelligent Automation and Computing 6(1):31–45
55. Udoh H (2024) E-governance performance in the context of developing countries. Doctoral dissertation, University of Leicester
56. Veloso Á, Fonseca F, Ramos R (2024) Insights from smart city initiatives for urban sustainability and contemporary urbanism. Smart Cities 7(6):3188–3209
57. Waddell S (2017) Societal learning and change: how governments, business and civil society are creating solutions to complex multi-stakeholder problems. Routledge, New York and London
58. Wang M, Zhou T (2023) Does smart city implementation improve the subjective quality of life? Evidence from China. *Technology in Society* 72:102161

Chapter 2
Bridging Innovation and Sustainability: Global Insights on ICT Applications

2.1 The Intersection of ICT and Sustainability

The convergence of ICT and sustainability is reshaping industries, governance, and society, creating a multidimensional nexus that influences environmental, economic, and social systems [20, 33]. This section explores the key intersections where ICT drives sustainability, focusing on four primary areas: smart infrastructure and urban resilience [19], resource efficiency and circular economy, climate change mitigation and adaptation, and social inclusion and digital equity.

ICT is a cornerstone of smart city development, where digital technologies enhance urban sustainability by improving efficiency, reducing waste, and fostering resilience. Smart infrastructure, including IoT-enabled transportation systems, AI-driven energy grids, and big data analytics for urban planning, helps cities optimize resource use and minimize environmental impact [65, 84]. For instance, smart grids use IoT and AI to balance electricity demand and supply, integrating renewable energy sources more effectively. These systems not only reduce carbon emissions but also enhance energy security and resilience against power disruptions. Similarly, real-time traffic management systems, powered by AI and data analytics, reduce congestion and pollution by optimizing vehicle flows, promoting public transportation, and integrating shared mobility solutions like electric bikes and autonomous shuttles.

Moreover, ICT enhances disaster resilience by enabling early warning systems and real-time monitoring of environmental risks [67], [74], [35, 43]. AI-powered climate models predict extreme weather events, while satellite-based remote sensing provides crucial data for flood and wildfire management. In highly urbanized areas vulnerable to climate change, these technologies help mitigate risks and ensure sustainable urban development.

One of the strongest intersections between ICT and sustainability lies in promoting a circular economy, where materials and products are reused, refurbished, or recycled to minimize waste and resource depletion. Digital solutions enable businesses and

A. Cheshmehzangi, *Urban Innovation and Sustainability*, Urban Sustainability,
https://doi.org/10.1007/978-981-96-6370-5_2

policymakers to track resource flows, reduce inefficiencies, and transition from linear to circular economic models. Blockchain technology, for example, enhances supply chain transparency by providing a tamper-proof record of material origins, production processes, and recycling practices [71, 79]. This ensures ethical sourcing, reduces waste, and helps companies meet sustainability targets. Additionally, AI-driven analytics optimize resource management by identifying patterns in consumption and suggesting sustainable alternatives.

E-waste management is another area where ICT plays a crucial role. As the digital economy expands, so does electronic waste, making responsible disposal and recycling critical [44, 45]. IoT-enabled smart waste management systems help monitor waste generation and streamline recycling efforts, ensuring that valuable materials [41], [46], such as rare earth metals, are recovered efficiently.

In addition, ICT solutions are vital in the global response to climate change, both in mitigating emissions and enhancing adaptation strategies [2]. Digital technologies contribute to decarbonization efforts by optimizing energy efficiency, promoting low-carbon industries, and enabling carbon footprint tracking [75]. For instance, cloud computing and AI facilitate climate modelling, helping scientists and policymakers predict long-term climate trends and devise mitigation strategies [72]. Digital platforms also support carbon trading markets by ensuring transparency in emissions reporting and offset mechanisms. Moreover, remote sensing technologies, including drones and satellites, monitor deforestation, ocean temperatures, and glacier movements [56, 69], providing real-time insights into environmental changes.

In agriculture, precision farming powered by IoT and AI enhances climate resilience by optimizing water usage, reducing pesticide dependency, and improving crop yields. Sensors detect soil moisture levels, allowing farmers to irrigate efficiently, while AI analyzes weather patterns to recommend adaptive farming techniques [7]. Such innovations reduce resource wastage and enhance food security in the face of climate uncertainties.

While ICT accelerates sustainability, its benefits must be equitably distributed to avoid exacerbating social inequalities [21]. Digital inclusion is critical in ensuring that marginalized communities—particularly in developing countries—have access to sustainable ICT-driven solutions [14, 38]. Bridging the digital divide enhances economic opportunities, social well-being, and climate resilience. For example, mobile banking and digital finance empower underbanked populations by providing access to microloans and digital payments [1]. This financial inclusion enables small-scale entrepreneurs, including women and rural farmers, to invest in sustainable livelihoods. Additionally, telemedicine and e-health solutions improve healthcare access in remote areas, reducing the need for travel and associated carbon emissions.

Open data platforms and civic tech initiatives also strengthen participatory governance by enabling communities to engage in decision-making processes [17, 76]. Digital tools allow citizens to monitor environmental issues, report pollution violations, and advocate for sustainable policies. When governments integrate ICT into sustainability governance, transparency and accountability improve (Singh and [80]), fostering more inclusive and effective policymaking.

The intersection of ICT and sustainability is vast and multidimensional, influencing how societies manage resources, adapt to climate change, and promote equity [47]. Digital technologies not only enable smarter, greener cities but also empower communities with information and tools to participate in sustainable development. However, realizing the full potential of ICT requires a balanced approach that addresses digital disparities, minimizes environmental footprints, and ensures ethical governance of technological advancements.

As ICT continues to evolve, its integration with sustainability efforts will shape the future of urban development, energy transitions, and social equity [4, 54]. The next sections of this chapter will explore specific applications of ICT in sustainability, the challenges associated with implementation, and the opportunities for leveraging digital innovations for a more sustainable and inclusive future.

2.1.1 Defining ICT in the Context of Sustainability

Nowadays, ICTs have become central to modern society, shaping economic systems, governance, and daily life. In the context of sustainability, ICT refers to digital technologies—including artificial intelligence (AI), the Internet of Things (IoT), blockchain, big data analytics, and cloud computing—that enhance environmental, social, and economic well-being. These technologies offer transformative solutions to some of the most pressing global challenges, including climate change, resource depletion, and social inequality.

ICT applications in sustainability extend across various sectors, from smart energy grids optimizing electricity consumption to precision agriculture reducing resource waste [62]. Digital platforms also play a role in promoting sustainable behaviour by providing individuals and organizations with real-time data on their ecological footprint [52]. For instance, IoT sensors in buildings can significantly reduce energy consumption, while blockchain technology can improve supply chain transparency for fair trade and ethical sourcing. These innovations demonstrate that ICT is not only an enabler of economic growth but also a critical tool in advancing sustainable development goals (SDGs).

2.1.2 The Role of Digital Innovation in Sustainable Development

The integration of ICT into sustainability efforts aligns with global development agendas, particularly the United Nations' 2030 Agenda for Sustainable Development. Digital solutions contribute to SDG targets such as affordable and clean energy (SDG 7), sustainable cities and communities (SDG 11), and climate action (SDG 13) [28].

For example, smart mobility systems reduce urban congestion and emissions, while AI-driven environmental monitoring enhances disaster preparedness and resilience.

ICT fosters innovation by enabling new business models and enhancing resource efficiency. The shift toward circular economies, where products and materials are reused and recycled, relies on digital tracking systems and AI-driven analytics to optimize material flows [16, 82, 83]. Moreover, digital finance, including mobile banking and blockchain-based transactions, expands financial inclusion, empowering marginalized communities with access to economic opportunities.

Despite these benefits, the integration of ICT in sustainability comes with challenges. Digital infrastructure requires significant energy and material resources, raising concerns about e-waste and carbon footprints [9]. Additionally, unequal access to ICT—known as the digital divide—can exacerbate social inequalities if not addressed through inclusive policies. Therefore, while ICT has immense potential to drive sustainability, its implementation must be carefully managed to balance innovation with responsible resource use and equitable access.

As digital technologies continue to evolve, their role in sustainability will expand, offering new pathways for addressing global environmental and social challenges. The following sections explore how ICT is applied across various sectors, the challenges it faces, and the opportunities for scaling sustainable digital solutions worldwide.

2.2 ICT Applications in Key Sustainability Sectors

ICTs have emerged as transformative tools in advancing sustainability across multiple sectors [10]. As societies increasingly grapple with environmental challenges, resource constraints, and rapid urbanization, digital innovations provide practical solutions to enhance resilience, efficiency, and sustainability. From improving urban management to mitigating climate change, ICT applications are reshaping industries, governance, and daily life.

Two critical areas where ICT plays a vital role are smart cities and urban resilience and digital solutions for climate action and resource efficiency. In the urban sector, smart technologies optimize energy use, reduce congestion, enhance public services, and support sustainable infrastructure [77]. Meanwhile, ICT-driven solutions for climate action enable carbon footprint tracking, enhance resource efficiency, and facilitate data-driven decision-making for environmental sustainability. This section explores how these ICT applications are driving transformative change and enabling the transition towards a more sustainable future.

2.2.1 Smart Cities and Urban Resilience

Cities all across the world are under tremendous pressure to change in ways that support resilience, sustainability, and efficiency due to the fast rate of urbanization [55]. Cities must come up with creative ways to deal with issues like resource consumption, waste management, transportation, and climate-related hazards as urban populations continue to rise. By optimizing resource use, improving infrastructure, boosting resilience against climate-related disasters, and improving urban planning, smart cities—which leverage the power of information and communication technologies, or ICT—offer revolutionary solutions to these problems.

Smart cities can create dynamic and responsive environments by combining technologies like cloud computing, big data analytics, artificial intelligence (AI), and the Internet of Things (IoT). Real-time monitoring and management of resources like waste, energy, and water is made possible by IoT devices integrated into city infrastructure. Cities are able to increase overall efficiency, minimize waste, and maximize consumption as a result. For instance, big data analytics assists urban planners in making data-informed decisions regarding future development, while AI-driven algorithms can analyze traffic patterns and modify signal timings to reduce congestion and emissions.

Additionally, smart cities increase resiliency to climate-related disasters. Cities can keep an eye on environmental hazards like flooding, heat waves, and air quality with the aid of remote sensing technologies, IoT-enabled sensors, and predictive analytics. With the use of these tools, authorities can minimize damage by implementing mitigation measures, optimizing emergency response plans, and issuing early warnings. Long-term sustainability objectives are also aided by the integration of renewable energy sources into urban infrastructure, such as solar-powered systems and smart grids, which lessen reliance on non-renewable resources.

Ultimately, the application of ICT in smart cities fosters a higher quality of life for urban residents by improving safety, mobility, and environmental sustainability, while also reducing the carbon footprint. These advancements not only improve the living conditions in urban areas but also position cities to thrive in the face of increasingly complex environmental and social challenges. The ability to adapt to these challenges and build resilient, sustainable cities is critical for shaping the future of urban living on a global scale.

2.2.1.1 ICT Applications in Smart Cities

Here, we summarize ICT applications in smart cities into four critical categories of (1) Smart Infrastructure and Energy Management, (2) Intelligent Transportation Systems (ITS), (3) Smart Water and Waste Management, and (4) Disaster Preparedness and Urban Resilience (Fig. 2.1).

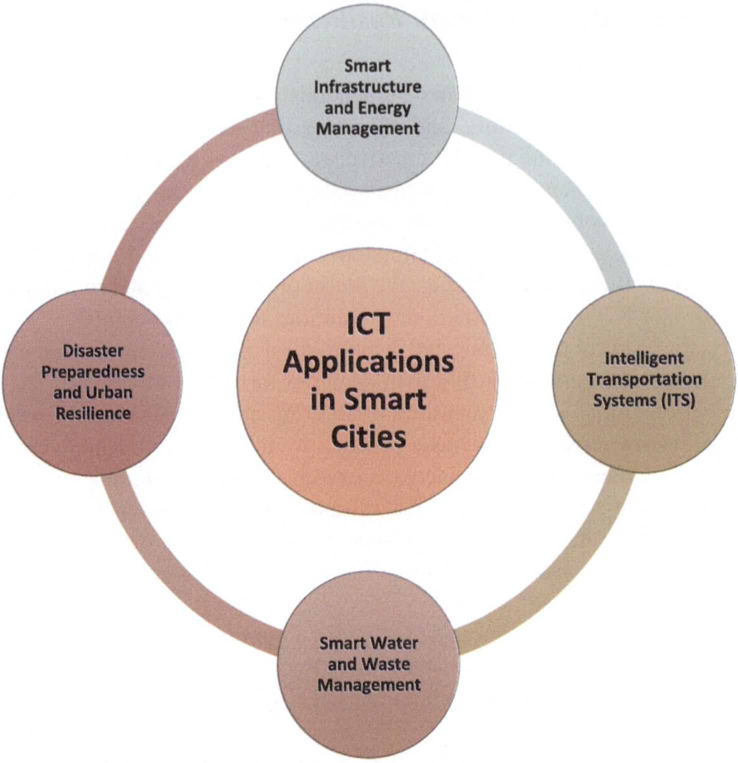

Fig. 2.1 Summary of four categories of 'ICT Applications in Smart Cities'

Smart Infrastructure and Energy Management

Energy management is one of the most important ICT applications in smart cities. Real-time electricity consumption monitoring by smart grids, which are fuelled by IoT sensors and AI-driven analytics, enables improved load balancing and the integration of renewable energy sources [32]. These grids enhance energy security, minimize energy waste, and lessen dependency on fossil fuels. Smart grid technologies, for instance, have been implemented in cities like Singapore and Amsterdam to optimize energy distribution and lower carbon emissions.

Intelligent Transportation Systems (ITS)

One of the main causes of traffic and emissions in cities is transportation. ICT-driven solutions that enhance mobility and lessen environmental effects include connected public transportation systems, adaptive signal control, and AI-powered traffic management. AI-enabled smart traffic lights, for instance, can modify signal timings in response to current traffic patterns, cutting down on emissions and idle time [42]. To improve the sustainability of urban mobility, ride-sharing services, and EV charging networks are also progressively incorporating ICT.

Smart Water and Waste Management

Efficient waste and water systems are also essential to sustainable urban management. IoT-enabled water monitoring systems optimize distribution networks, stop water waste, and find leaks. Sensor-equipped smart bins in waste management notify authorities when they are full, cutting down on pointless collection trips and increasing recycling effectiveness [18]. Smart waste collection systems have been put in place in cities like Seoul and Barcelona to expedite waste disposal and advance the ideas of the circular economy.

Disaster Preparedness and Urban Resilience

Urban resilience is important because natural disasters are occurring more frequently and with greater intensity due to climate change. Through emergency response systems, real-time monitoring, and predictive analytics, ICT is essential to disaster risk management. While GIS-based mapping systems facilitate improved disaster preparedness, AI-driven climate models aid in the prediction of extreme weather events [8]. Japan's earthquake early warning system, for instance, uses AI and Internet of Things sensors to identify seismic activity and send out alerts in a matter of seconds, allowing authorities and citizens to take appropriate action. In conclusion, ICT integration makes smart cities more resilient and sustainable over the long run by making them more efficient, adaptable, and ecologically friendly.

2.2.2 Digital Solutions for Climate Action and Resource Efficiency

Among the twenty-first century's most urgent sustainability issues are combating climate change and increasing resource efficiency. There has never been a more pressing need to find creative ways to lessen environmental impacts as the world's population grows and the demand on natural resources increases. By providing a variety of instruments and tactics that improve environmental sustainability across multiple industries, information and communication technologies, or ICT, are playing a crucial part in tackling these issues [10]. We can enhance energy efficiency, encourage a more sustainable use of resources, and lessen the negative effects of climate change by integrating digital technologies.

The ability of ICT to facilitate precise carbon footprint tracking is one of its primary contributions to climate action. Businesses, governments, and individuals can track and measure their carbon emissions in real-time using digital platforms and software tools, which provide vital information that can guide climate action plans. Organizations can use this information to make better decisions about how to cut emissions, increase energy efficiency, and meet carbon reduction targets [37]. Additionally, ICT plays a key role in advancing energy efficiency through the implementation of smart grids, which balance energy supply and demand to minimize waste, optimize electricity distribution, and integrate renewable energy sources.

Along with energy efficiency, ICT is revolutionizing supply chains and advancing sustainability through waste reduction and transparency [6]. Businesses can track the origins of raw materials, keep an eye on production procedures, and make sure supply chains aren't causing environmental damage by utilizing technologies like blockchain. ICT can help minimize waste, cut down on inefficiencies, and guarantee ethical sourcing methods that support sustainability objectives by increasing supply chain visibility.

Additionally, ICT facilitates the shift to a circular economy by making digital platforms for waste recycling, product life-cycle management, and resource tracking possible. Businesses can decrease waste production, increase product longevity, and lessen their reliance on virgin resources by employing digital tools to track material usage, reuse, and recycling [50]. This lessens the environmental impact of conventional linear economic models and promotes more sustainable consumption patterns. Ultimately, the integration of ICT in climate action and resource efficiency fosters better decision-making, enhances accountability, and accelerates global efforts to combat climate change. By providing actionable insights, driving innovation, and supporting collaborative solutions, ICT plays a critical role in building a more sustainable and resilient global economy.

2.2.2.1 ICT Applications in Climate Action

Here, we summarize ICT applications in climate action into four critical categories of: (1) AI and Big Data for Carbon Footprint Reduction, (2) Blockchain for Transparent and Sustainable Supply Chains, (3) IoT for Smart Energy Management, and (4) Climate Monitoring and Early Warning Systems (Fig. 2.2).

AI and Big Data for Carbon Footprint Reduction

Big data analytics and artificial intelligence offer strong instruments for quantifying and lowering carbon footprints. In order to identify inefficiencies in energy consumption, transportation networks, and industrial processes, artificial intelligence (AI) algorithms examine enormous datasets and provide suggestions for carbon reduction tactics [66]. For instance, Google's AI-powered carbon footprint tracker assists

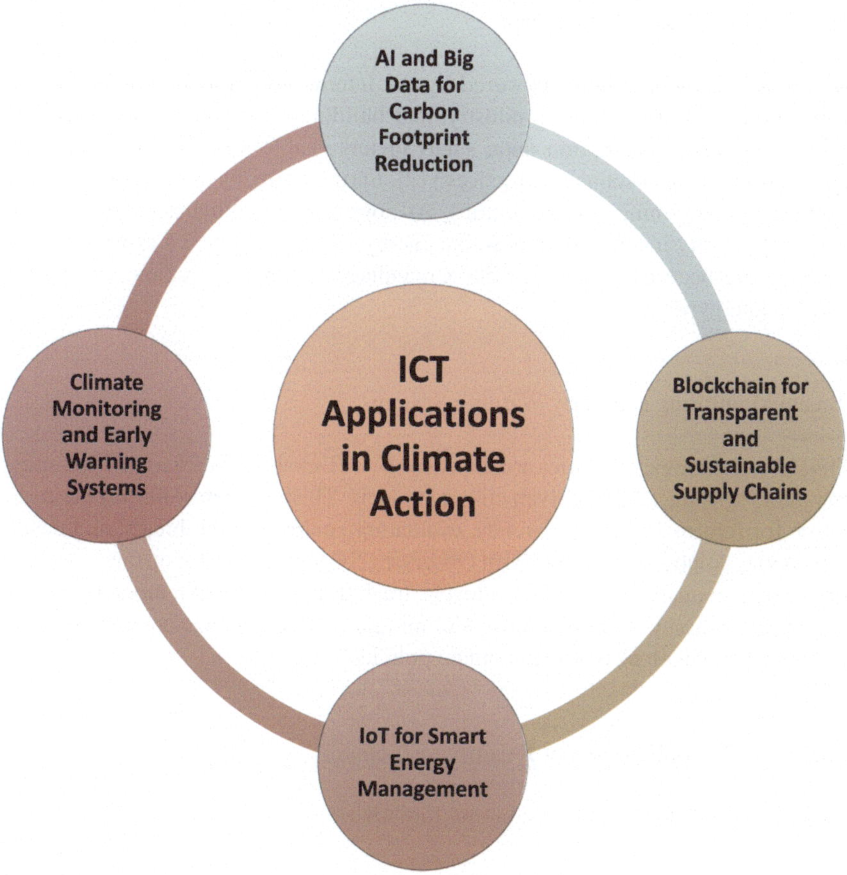

Fig. 2.2 Summary of four categories of 'ICT Applications in Climate Action'

companies in tracking and reducing their emissions, and satellite monitoring powered by AI identifies illicit mining and deforestation.

Blockchain for Transparent and Sustainable Supply Chains

Blockchain technology improves supply chain transparency and traceability, guaranteeing that goods fulfill sustainability requirements [22]. Blockchain can be used by businesses to monitor carbon emissions, confirm ethical sourcing, and stop greenwashing. For example, retailers can track food products from farm to shelf using IBM's blockchain-based Food Trust network, which ensures sustainable sourcing and reduces food waste.

IoT for Smart Energy Management

Energy management systems powered by the Internet of Things optimize the use of electricity in transportation, industry, and buildings. By giving users access to real-time energy consumption data, smart meters enable them to spot inefficiencies and implement sustainable practices [12]. Furthermore, by anticipating demand trends and streamlining storage options, AI-powered energy management systems more successfully incorporate renewable energy sources. IoT-based district heating systems have been used by cities like Copenhagen to increase efficiency and cut down on energy waste.

Climate Monitoring and Early Warning Systems

Climate monitoring capabilities are improved by IoT-enabled weather stations, satellite remote sensing, and AI-driven climate models. These systems offer useful information for forecasting climate trends, evaluating environmental threats, and developing policy solutions. NASA's Earth Observing System (EOS), for instance, gathers climate data to analyze air pollution levels, track trends in global temperatures, and track deforestation [39]. Global efforts to mitigate climate change are supported by these insights, which also inform climate policies.

2.2.2.2 ICT Applications in Resource Efficiency

Digital Platforms for Circular Economy Initiatives

By encouraging waste reduction and resource efficiency, ICT is transforming the shift to a circular economy. Digital platforms make it easier to track the lifecycle of products, encourage sharing economies, and allow for the reuse of materials. Platforms such as Too Good To Go, for instance, link companies with excess food to customers at a discount, lowering food waste and enhancing food security [11]. Similarly, by locating and separating valuable materials from waste streams, AI-driven material recovery platforms maximize recycling procedures.

Smart Agriculture for Sustainable Resource Use

Precision farming technologies use drones, IoT, and AI to improve resource efficiency in the agricultural industry. Farmers can optimize irrigation and minimize water waste by using real-time moisture level data from IoT-enabled soil sensors [49]. Analytics driven by AI forecast weather trends and suggest adaptive farming methods, increasing output while reducing environmental effects. Smart agriculture techniques have been pioneered by nations like the Netherlands [5, 13], which use AI to increase crop yields while using less fertilizer and pesticide.

Digital Twin Technology for Resource Optimization

Real-time monitoring and simulation of resource use in cities and industries are made possible by digital twins, which are virtual copies of physical assets. Stakeholders can maximize energy use, cut waste, and improve efficiency by examining digital models of factories, buildings, and urban infrastructure [27]. Digital twin technology aids engineers and architects in the construction industry in creating carbon footprint-reducing, energy-efficient structures.

Finally, we contend that ICT applications are leading the way in promoting climate action and sustainability in urban development. While digital innovations in climate action enable carbon reduction, promote circular economy practices, and improve resource efficiency, smart city technologies improve urban resilience, optimize infrastructure, and improve quality of life. To guarantee inclusive and long-lasting ICT adoption, however, issues like cybersecurity, data privacy, and digital equity must be resolved. Going forward, attaining long-term environmental, social, and economic objectives will require utilizing ICT for sustainability.

2.3 Challenges and Opportunities in Implementing ICTs for Sustainability

ICT integration into sustainability initiatives has enormous potential to address global issues like resource depletion, social injustice, and climate change [51]. However, there are challenges to the broad use of ICT for sustainability. There are many obstacles to overcome, including the digital divide, complicated regulations, limited infrastructure, and environmental concerns. These difficulties also offer chances for policy changes, inclusive development, and innovation [28]. Effective use of ICTs to promote sustainable development requires an understanding of the ratio of enablers to barriers.

Even though digital technologies have fuelled economic growth, not everyone has benefited equally from them. Inequitable participation in sustainability initiatives is still hampered by the digital divide [29–31], which is the difference between those who have and do not have access to ICT. Inequalities in education, economic opportunity, and climate resilience result from the lack of infrastructure and resources in many rural and marginalized communities. Concerns are also raised by the growth of ICT infrastructure regarding its effects on the environment, including energy use, the production of e-waste, and the extraction of resources for the production of digital devices.

Policies and regulations that direct the adoption of ICT in a sustainable manner are shaped in large part by governments and international organizations [36]. For ICTs to maximize their benefits while minimizing their risks, policy frameworks that support digital inclusion, cybersecurity, data privacy, and sustainable technology development are crucial. Through the examination of international case studies, this section looks at the opportunities and difficulties of using ICTs for sustainability, with an emphasis on digital equity and accessibility as well as the policy frameworks that support responsible and efficient ICT adoption.

2.3.1 Bridging the Digital Divide: Accessibility and Equity Issues

ICT is developing at a rapid pace, but digital accessibility is still a major global concern. Rural and impoverished communities around the world often face challenges in obtaining digital devices, dependable internet, and critical ICT-based services. Existing socioeconomic disparities are made worse by the digital divide, especially in the areas of economic opportunity, healthcare, and education. Marginalized groups run the risk of losing out on the advantages of digital sustainability solutions in the absence of focused interventions.

In sub-Saharan Africa, internet penetration is still much lower than in developed regions, making it one of the most notable instances of the digital divide [59]. Rural communities frequently lack the infrastructure required for ICT access, despite the rapid digitization of urban centers in nations like South Africa and Kenya [64]. This gap is a systemic problem connected to economic inequality, governance constraints, and deficiencies in energy infrastructure rather than just a technical one. Even in areas with inadequate banking infrastructure, ICT can close financial gaps, as shown by programs like Kenya's M-Pesa mobile banking platform [53]. By giving millions of people access to digital financial services, M-Pesa has increased economic inclusion and decreased poverty, especially in rural areas.

Another demonstration is India's Digital India initiative, which seeks to advance digital literacy and e-governance while giving internet access to marginalized communities [68]. The Indian government has brought broadband connectivity to thousands of rural villages through projects like the BharatNet project, opening up access to online marketplaces, telemedicine, and e-learning. India is tackling digital exclusion and generating prospects for sustainable economic growth by incorporating ICT into a number of sectors. But there are still issues, especially with addressing gender differences in ICT use and guaranteeing inexpensive access to digital services.

The digital divide is particularly noticeable in remote and indigenous communities in Latin America, where access to the internet is restricted by financial and geographic limitations [73]. By implementing cutting-edge connectivity solutions, groups like Peru's Internet para Todos (Internet for All) initiative are attempting to bridge this divide. The initiative has made digital education, telehealth, and agricultural technologies accessible to rural populations through inexpensive, solar-powered internet infrastructure. This illustrates how ICT solutions driven by sustainability can be customized to fit particular environmental and socioeconomic circumstances.

International organizations, the private sector, and governments must work together to close the digital divide. To guarantee that ICT solutions support inclusive and sustainable development, investments in digital infrastructure, financial aid for reasonably priced devices, and digital literacy initiatives are crucial. Furthermore, by utilizing the resources and experience of tech firms to increase connectivity in underserved areas, public–private partnerships can hasten development.

2.3.2 Policy Frameworks and Governance for Sustainable ICT Adoption

Effective governance and policy frameworks, in addition to technological advancements, are essential for the successful application of ICTs for sustainability. Policies that guarantee the responsible and fair use of digital technologies are largely shaped by governments and regulatory agencies. To promote sustainable ICT adoption, policies that address cybersecurity, data privacy, environmental effects, and ethical issues are required. Finding a balance between innovation and regulatory oversight is one of the main issues in ICT governance. The General Data Protection Regulation (GDPR) of the European Union has established a global standard for data privacy, impacting the way personal data is handled in digital sustainability initiatives [86]. For companies and organizations putting digital sustainability solutions into practice, GDPR poses compliance issues even though it guarantees accountability and transparency in data usage. Fostering responsible digital transformation requires finding a balance between technological innovation and data protection.

Through programs like the Smart Cities initiative, the Chinese government has actively shaped ICT-driven sustainability policies. AI-powered urban management systems have been deployed by cities like Shenzhen and Hangzhou to improve public services, minimize energy use, and optimize traffic flow [48]. These initiatives show how ICT adoption influenced by policy can support social and environmental sustainability. However, worries about government monitoring and data privacy emphasize how crucial it is to create legal frameworks that uphold citizens' rights while utilizing digital innovations for sustainability.

Policies addressing e-waste management, energy-efficient ICT infrastructure, and responsible material sourcing are necessary for the shift to a circular digital economy. Extended producer responsibility, repairability standards, and sustainable ICT manufacturing are all covered in the European Commission's Circular Economy Action Plan. Policymakers can reduce the environmental impact of ICT while fostering resource efficiency by providing incentives to businesses that create environmentally friendly digital products.

Partnerships between the public and private sectors are also crucial in promoting innovative policies for long-term ICT adoption. A pan-African coalition of governments and tech firms, Smart Africa aims to establish a uniform framework for digital policies throughout African countries [61]. Smart Africa seeks to improve digital connectivity, economic integration, and sustainable development by standardizing regulations and encouraging cross-border ICT investments [34]. These programs serve as examples of how governance frameworks can encourage the adoption of ICT while guaranteeing that new developments in technology are in line with sustainability objectives.

Federal and state governments in the US have implemented measures to encourage green ICT projects, such as financing for smart grid technologies and tax breaks for energy-efficient data centers. For instance, rules requiring data centers to implement sustainable energy practices have been put into place by the California Energy Commission [3]. These regulations address issues with energy use and carbon emissions while emphasizing the role of government action in directing ICT development toward sustainability. The use of blockchain technology for accountability and transparency is one of the most exciting prospects in ICT governance [63]. Because blockchain-based governance models offer unchangeable records of environmental data, carbon credits, and ethical supply chains, they can increase confidence in digital sustainability initiatives [70]. To improve transparency and cut down on bureaucratic inefficiencies, Estonia, a leader in digital governance worldwide, has adopted blockchain-based public services [24]. By applying this model to sustainability initiatives, digital solutions can be made to be resilient, secure, and accountable.

Policymakers must implement adaptive governance techniques that keep up with technological breakthroughs as the digital landscape continues to change. A framework for evaluating the sustainability impact of new ICT innovations prior to their widespread adoption is provided by regulatory sandboxes, where cutting-edge technologies are tested in controlled environments [15]. Governments can reduce potential risks and create an environment that supports ICT-driven sustainability by encouraging regulatory agility. Digital exclusion, complicated regulations, and environmental concerns are some of the difficulties that come with using ICTs for sustainability, but there are also a lot of chances for innovation, inclusiveness, and responsible governance. Investing in connectivity, digital literacy, and inclusive policies is necessary to close the digital divide and guarantee that underserved communities gain from technological advancements. Simultaneously, governance mechanisms and policy frameworks are essential for directing the sustainable adoption of ICT [25], striking a balance between advancements in technology and ethical, environmental, and security concerns.

Stakeholders can create ICT solutions that are not only technologically sophisticated but also socially and environmentally responsible by studying international case studies and encouraging international collaboration [40]. In the future, a cooperative, policy-driven strategy that guarantees fair access, regulatory integrity, and long-term environmental resilience will be necessary for the convergence of digital technologies and sustainability initiatives.

2.4 Future Directions: Strengthening ICT's Role in Global Sustainability

The importance of information and communication technologies (ICTs) in promoting revolutionary solutions is growing as the globe confronts more and more sustainability issues. ICT applications, from resource efficiency to climate resilience, are influencing sustainable development in the future by facilitating more informed decision-making, maximizing energy use, and promoting equitable economic growth [4]. However, realizing the full potential of digital technologies necessitates a forward-thinking strategy that incorporates cutting-edge technologies like blockchain, the Internet of Things (IoT), and artificial intelligence (AI) while encouraging international cooperation and knowledge-sharing to scale sustainable ICT solutions.

Using AI to improve sustainability results is one of the most exciting future directions. Large volumes of environmental data can be processed by AI-powered analytics to forecast climate trends, optimize energy grids, and boost agricultural output. For instance, governments and conservation groups are able to take preventative action by using machine learning models to evaluate satellite imagery and identify deforestation trends in real time. Similarly, by dynamically modifying energy use in response to demand patterns, AI-driven energy management systems are assisting cities in lowering carbon emissions [58, 60, 81]. With intelligent algorithms improving air quality monitoring, lowering traffic, and optimizing public transportation routes, the use of AI in sustainable urban planning is also gaining traction.

IoT, in conjunction with AI, is transforming sustainability initiatives by offering automation and real-time monitoring across industries. Smart sensors that are integrated into industrial, agricultural, and infrastructure processes allow for accurate resource management, which lowers waste and boosts productivity [78]. IoT-enabled irrigation systems ensure optimal water usage in water-scarce regions by adjusting according to weather forecasts and soil moisture levels. IoT devices are enabling intelligent waste management in smart cities. Sensors in these devices identify when bins are full and optimize collection routes to reduce emissions and fuel consumption. Sustainability initiatives can be made even more successful by combining IoT and AI, which enables predictive and adaptive interventions that improve resilience.

Another significant new development that has the potential to improve transparency and sustainability governance is blockchain technology. Blockchain improves accountability and trust in carbon trading, ethical supply chains, and environmental monitoring by generating unchangeable digital records [87]. To make sure that products fulfil environmental and social responsibility standards, blockchain-based platforms, for example, are being used to track the origin and sustainability credentials of raw materials. Blockchain enables peer-to-peer energy trading in the renewable energy space [85], enabling solar-equipped homes to decentralizedly sell extra energy to their neighbours. Lowering dependency on centralized power grids, not only encourages the use of renewable energy but also democratizes energy access.

Notwithstanding these technological developments, the shift to global sustainability cannot be accomplished in a vacuum. Strong international collaboration, knowledge-sharing systems, and cross-sector alliances are necessary to enhance ICT's contribution to sustainability. Because of their complexity, sustainability challenges necessitate cooperative solutions that incorporate knowledge from businesses, academia, governments, and civil society [57]. Global frameworks like the Sustainable Development Goals (SDGs) of the UN are crucial for establishing international collaborations and guaranteeing that ICT advancements are in line with more general sustainability goals.

2.4.1 Concluding Remarks

The requirement for scalable and flexible ICT solutions that can be adjusted to various socioeconomic and environmental contexts is a critical component of international cooperation. High-income nations might be able to invest in advanced smart city infrastructures, but developing countries frequently struggle with funding, digital literacy, and policy support [23]. International organizations and tech companies must collaborate to create affordable, open-source, and interoperable solutions that can be used in various geographical locations to close this gap. Projects like the Global Partnership for Sustainable Development Data are essential in increasing the accessibility of data about sustainability, which empowers communities and governments to make defensible decisions.

Furthermore, to guarantee that innovations in digital sustainability are extensively shared and replicated, knowledge-sharing networks need to be reinforced. It is important to record and disseminate best practices from digital agriculture initiatives, smart city projects, and circular economy models. Empowering local communities to use digital technologies for sustainability requires capacity-building initiatives, such as training in digital skills and ICT policy advisory services [26]. Initiatives like ASEAN's Smart Cities Network and the African Union's Digital Transformation Strategy serve as prime examples of how regional collaboration can propel ICT-enabled sustainability solutions that are suited to particular cultural and economic contexts.

Harmonization of regulations and ethical considerations are also critical to the future of ICT-driven sustainability. As blockchain, IoT, and AI are incorporated more and more into sustainability initiatives, policymakers need to create governance frameworks that protect cybersecurity, data privacy, and moral AI applications. In the future, incorporating ICTs into sustainability plans will necessitate constant adjustment to new developments in technology and changing global issues. Future regulations ought to support ethical AI, sustainable ICT production, and digital product incentives based on the circular economy. To guarantee that ICT-driven sustainability solutions are inclusive, moral, and ecologically conscious, governments, corporations, and civil society organizations must collaborate.

Lastly, ICT's role in sustainability will depend on its capacity to bring about systemic, scalable change via creativity, fair access, and international collaboration. The world can use digital transformation to promote sustainable development and make sure that advancements in technology benefit both people and the environment by utilizing new technologies and bolstering cooperative frameworks.

References

1. Adelaja AO, Umeorah SC, Abikoye BE, Nezianya MC (2024) Advancing financial inclusion through fintech: solutions for unbanked and underbanked populations. World Journal of Advanced Research and Reviews 23(01):427–438
2. Adenle AA, Azadi H, Arbiol J (2015) Global assessment of technological innovation for climate change adaptation and mitigation in developing world. J Environ Manage 161:261–275
3. Adler EN (2017) Before the public utilities commission of The State of California, Rulemaking 13-11-005, Available from: https://eecoordinator.info/wp-content/uploads/2017/05/FINAL-SDGE-2016-EE-Annual-Report-with-attachment-.pdf
4. Agboola OP, Tunay M (2023) Urban resilience in the digital age: the influence of information-communication technology for sustainability. J Clean Prod 428:139304
5. Ahmad L, Nabi F (2021) Agriculture 5.0: artificial intelligence, IoT and machine learning. New York: CRC Press
6. Akbari M (2024) Revolutionizing supply chain and circular economy with edge computing: Systematic review, research themes and future directions. Manag Decis 62(9):2875–2899
7. Akintuyi OB (2024) Adaptive AI in precision agriculture: a review: investigating the use of self-learning algorithms in optimizing farm operations based on real-time data. Research Journal of Multidisciplinary Studies 7(02):016–030
8. Albahri AS, Khaleel YL, Habeeb MA, Ismael RD, Hameed QA, Deveci M, Homod RZ, Albahri OS, Alamoodi AH, Alzubaidi L (2024) A systematic review of trustworthy artificial intelligence applications in natural disasters. Comput Electr Eng 118:109409
9. Albreem MA, Sheikh AM, Bashir MJ, El-Saleh AA (2023) Towards green Internet of Things (IoT) for a sustainable future in Gulf cooperation council countries: current practices, challenges and future prospective. Wireless Netw 29(2):539–567
10. Andreopoulou ZS (2012) Green informatics: ICT for green and sustainability. J Agric Inform 3(2)
11. Apostolidis C, Brown D, Wijetunga D, Kathriarachchi E (2021) Sustainable value co-creation at the bottom of the Pyramid: using mobile applications to reduce food waste and improve food security. J Mark Manag 37(9–10):856–886
12. Avancini DB, Rodrigues JJ, Martins SG, Rabêlo RA, Al-Muhtadi J, Solic P (2019) Energy meters evolution in smart grids: a review. J Clean Prod 217:702–715
13. Ayaz M, Ammad-Uddin M, Sharif Z, Mansour A, Aggoune EHM (2019) Internet-of-Things (IoT)-based smart agriculture: toward making the fields talk. IEEE access 7:129551–129583
14. Aziz A (2020) Digital inclusion challenges in Bangladesh: the case of the National ICT Policy. Contemporary South Asia 28(3):304–319
15. Baldini D (2025) Legislative intersection perspectives on regulatory sandboxes: navigating the interplay between the AI Act and the GDPR. In: White Paper on Regulatory Sandboxes for AI and Cybersecurity. Cybersecurity National Lab
16. Bashynska I, Prokopenko O (2024) Innovative technologies and digital platforms: AI's role in a sustainable circular economy. Sci J Bielsk-Biala Sch Financ Law 28(3)
17. Bhanye J, Shayamunda R (2025) The promise of civic-tech: digital technologies and transparent, accountable governance. In: Digitalisation and public policy in Africa: GovTech and CivicTech innovations. Cham: Springer Nature Switzerland, pp 93–122

18. Biswas B, Imran MAU, Azizr MB, Noor SK, Ferdousmou J (2021) IoT technologies application and effect on urban infrastructure: intelligent waste management, energy efficiency, and traffic management. Journal of Environmental and Agricultural Studies 2(2):125–137
19. Caprotti F, Cowley R, Bailey I, Joss S, Sengers F, Raven R, Spaeth P, Jolivet E, Tan-Mullins M, Cheshmehzangi A, Xie L (2017) Smart eco-city development in Europe and China: policy directions. Exeter: University of Exeter, Available from: https://eprints.gla.ac.uk/182634/
20. Castro GDR, Fernandez MCG, Colsa AU (2021) Unleashing the convergence amid digitalization and sustainability towards pursuing the Sustainable Development Goals (SDGs): A holistic review. J Clean Prod 280:122204
21. Chancel L (2020) Unsustainable inequalities: social justice and the environment. Harvard University Press, Cambridge, MA
22. Chandan A, John M, Potdar V (2023) Achieving UN SDGs in food supply chain using blockchain technology. Sustainability 15(3):2109
23. Cheshmehzangi A (2016) City enhancement beyond the notion of "Sustainable City": Introduction to integrated assessment for city enhancement (iACE) toolkit. Energy Procedia 104:153–158
24. Cheshmehzangi A (2022) The impacts of ICT platforms on public place use in China. Built Environment 48(2):264–279
25. Cheshmehzangi A (2022b) The role of ICT and smartness for peace-building processes and practices. In ICT, cities, and reaching positive peace. Singapore: Springer Nature Singapore, pp 17–46
26. Cheshmehzangi A (2022) ICT, cities, and reaching positive peace. Springer, Singapore
27. Cheshmehzangi A, Pourroostaei Ardakani S (2024) Navigating the Future with the Nexus Between Digital Twin Computing and Urban Intelligence. In: Pourroostaei Ardakani S, Cheshmehzangi A (eds). Digital twin computing for urban intelligence. Singapore: Springer Nature Singapore, pp 245–256
28. Cheshmehzangi A, Zou T (2024) Literary analysis of the connection between urban innovations and Sustainable Development Goals (SDGs). Sustainability 16(15):6405
29. Cheshmehzangi A, Zou T, Su Z (2022) The digital divide impacts on mental health during the COVID-19 pandemic. Brain Behav Immun 101:211–213
30. Cheshmehzangi A, Zou T, Su Z, Tang T (2023) The growing digital divide in education among primary and secondary children during the COVID-19 pandemic: An overview of social exclusion and education equality issues. J Hum Behav Soc Environ 33(3):434–449
31. Cheshmehzangi A, Zou T, Su Z, Dawodu A, Ardakani SP, Sedrez M, Tang T (2024) Growing digital divide for expatriates population in China: a user perspective analysis During the COVID-19 Pandemic. SAGE Open 14(4):21582440241274612
32. Dahmani S (2024) Energy optimization and smart grids: IoT-based smart grid solution and smart grids applications. In: Harnessing high-performance computing and AI for environmental sustainability. IGI Global, pp 278–304
33. Elliot S (2011) Transdisciplinary perspectives on environmental sustainability: a resource base and framework for IT-enabled business transformation. Mis quarterly: 197–236
34. Elnahrawi N (2022) The role of digital transformation in Africa's regional integration. Available at SSRN 4595317
35. Esposito M, Palma L, Belli A, Sabbatini L, Pierleoni P (2022) Recent advances in internet of things solutions for early warning systems: a review. Sensors 22(6):2124
36. Faroqi H, Irajifar L, Cheshmehzangi A (2022) Sustainable development in smart and resilient local government areas: an empirical investigation of Victoria, Australia. Int J Sustain Dev & Plan 17(6)
37. Finnerty N, Sterling R, Contreras S, Coakley D, Keane MM (2018) Defining corporate energy policy and strategy to achieve carbon emissions reduction targets via energy management in non-energy intensive multi-site manufacturing organisations. Energy 151:913–929
38. Goel RK, Vishnoi S (2022) Urbanization and sustainable development for inclusiveness using ICTs. Telecommunications Policy 46(6):102311

39. Guo H, Fu W, Liu G (2019) Scientific satellite and moon-based earth observation for global change. Springer Singapore
40. Hosman L, Fife E (2008) Improving the prospects for sustainable ICT projects in the developing world. International journal of media & cultural politics 4(1):51–69
41. Kasulla S, Malik SJ, Baxla SP, Zafar S (2024) The role of Iot in waste management and sustainability. Partners Universal International Research Journal 3(2):76–88
42. Khan MW, Saad S, Ammad S, Rasheed K, Jamal Q (2024) Smart infrastructure and AI. In: AI in Material Science. CRC Press, pp. 193–215
43. Kirpalani C (2024) Technology-driven approaches to enhance disaster response and recovery. Geospatial Technol Nat Resour Manag 25–81
44. Kyriakopoulos GL (2022, October) The contribution of Digital Transformation Industry (DTI) in micro-and macro-economy. In: International scientific conference on Digital Transformation in Industry: trends, Management, Strategies. Cham: Springer Nature Switzerland, pp 101–108
45. Li S, Yang H, Huang W, Yang H (2025) Study on the dynamic coupled coordination and influencing factors of the digital economy, energy and human habitat in Western China. Sustainability Accounting, Management and Policy Journal. Available from: https://www.emerald.com/insight/content/doi/10.1108/sampj-11-2023-0826/full/html
46. Lakhouit A (2025) Revolutionizing urban solid waste management with AI and IoT: a review of smart solutions for waste collection, sorting, and recycling. Results in Engineering, 104018
47. Leach M, Reyers B, Bai X, Brondizio ES, Cook C, Díaz S, Espindola G, Scobie M, Stafford-Smith M, Subramanian SM (2018) Equity and sustainability in the Anthropocene: A social–ecological systems perspective on their intertwined futures. Global Sustainability 1:e13
48. Li G, Wang J, Wang X (2023) Construction and path of urban public safety governance and crisis management optimization model integrating artificial intelligence technology. Sustainability 15(9):7487
49. Liang C, Shah T (2023) IoT in agriculture: the future of precision monitoring and data-driven farming. Eigenpub Review of Science and Technology 7(1):85–104
50. Maiurova A, Kurniawan TA, Kustikova M, Bykovskaia E, Othman MHD, Singh D, Goh HH (2022) Promoting digital transformation in waste collection service and waste recycling in Moscow (Russia): applying a circular economy paradigm to mitigate climate change impacts on the environment. J Clean Prod 354:131604
51. Makrakis V (2017) Unlocking the potentiality and actuality of ICTs in developing sustainability-justice curricula and society. Knowl Cult 5(02):103–122
52. Malik S (2024) Data-driven decision-making: leveraging the IoT for real-time sustainability in organizational behavior. Sustainability 16(15):6302
53. Mbiti I, Weil DN (2015) Mobile banking: the impact of M-Pesa in Kenya. African successes, Volume III: Modernization and development. University of Chicago Press, Chicago, pp 247–293
54. McCormick K, Anderberg S, Coenen L, Neij L (2013) Advancing sustainable urban transformation. J Clean Prod 50:1–11
55. Mendizabal M, Heidrich O, Feliu E, García-Blanco G, Mendizabal A (2018) Stimulating urban transition and transformation to achieve sustainable and resilient cities. Renew Sustain Energy Rev 94:410–418
56. Mertikas SP, Partsinevelos P, Mavrocordatos C, Maximenko NA (2021) Environmental applications of remote sensing. In: Pollution assessment for sustainable practices in applied sciences and engineering. Butterworth-Heinemann, pp 107–163
57. Modara M (2019) The influence of government-private sector collaboration on innovation in a developing knowledge economy: the case of Bahrain. Doctoral dissertation. Available online from: http://dspace.bu.ac.th/handle/123456789/4122
58. Muniandi B, Maurya PK, Bhavani CH, Kulkarni S, Yellu RR, Chauhan N (2024) AI-driven energy management systems for smart buildings. Power System Technology 48(1):322–337
59. Mutula SM (2005) Peculiarities of the digital divide in sub-Saharan Africa. Program 39(2):122–138

60. Ning MA (2024) Artificial intelligence-driven decision support systems for sustainable energy management in smart cities. Int J Adv Comput Sci & Appl 15(9)
61. Nyirenda-Jere T, Mangwende T, Nimako G, Gashami JP, Samatar OE, ATU MS, Seck M (2022) African Union (AU) data policy framework. Policy. Available from: https://dig.watch/resource/african-union-au-data-policy-framework
62. Odara S, Khan Z, Ustun, TS (2015, July) Integration of precision agriculture and smartgrid technologies for sustainable development. In: 2015 IEEE Technological Innovation in ICT for Agriculture and Rural Development (TIAR). IEEE, pp. 84–89
63. Oliveira TA, Oliver M, Ramalhinho H (2020) Challenges for connecting citizens and smart cities: ICT, e-governance and blockchain. Sustainability 12(7):2926
64. Omweri FS (2024) A systematic literature review of e-government implementation in developing countries: examining urban-rural disparities, institutional capacity, and socio-cultural factors in the context of local governance and progress towards SDG 16.6. Int J Res Innov Soc Sci 8(8):1173–1199
65. Padhiary M, Roy P, Roy D (2025) The future of urban connectivity: Ai and iot in smart cities. In: Sustainable smart cities and the future of urban development. IGI Global Scientific Publishing, pp 33–66
66. Pimenow S, Pimenowa O, Prus P (2024) Challenges of artificial intelligence development in the context of energy consumption and impact on climate change. Energies 17(23):5965
67. Poslad S, Middleton SE, Chaves F, Tao R, Necmioglu O, Bügel U (2015) A semantic IoT early warning system for natural environment crisis management. IEEE Trans Emerg Top Comput 3(2):246–257
68. Prasad K (2012) E-governance policy for modernizing government through digital democracy in India. J Inf Policy 2:183–203
69. Purkis SJ, Klemas VV (2011) Remote sensing and global environmental change. John Wiley & Sons, New Jersey
70. Rani P, Sharma P, Gupta I (2024) Toward a greener future: A survey on sustainable blockchain applications and impact. J Environ Manage 354:120273
71. Raslan A, Cheshmehzangi A, Towey D, Tizani, W, Kapogiannis G (2024) Impact factor analysis affecting the operation of asset lifecycle by using asset information modeling and blockchain technology. Construction Innovation. Available from: https://www.emerald.com/insight/content/doi/10.1108/CI-10-2022-0272/full/html
72. Reddy MS, Reddy CKK, Hanafiah MM (2024) Climate change mitigation and adaptation strategies enhanced by intelligent systems in Industry 6.0. In: Maintaining a sustainable world in the nexus of environmental science and AI. IGI Global, pp 201–228
73. Robinson L, Schulz J, Dodel M, Correa T, Villanueva-Mansilla E, Leal S, Magallanes-Blanco C, Rodriguez-Medina L, Dunn HS, Levine L, McMahon R (2020) Digital inclusion across the Americas and Caribbean. Social Inclusion 8(2):244–259
74. Sarker MNI, Peng Y, Yiran C, Shouse RC (2020) Disaster resilience through big data: way to environmental sustainability. International Journal of Disaster Risk Reduction 51:101769
75. Sathianarayanan M, Kirdemir U, Gianoli A (2025) Urban intelligence: navigating climate challenges with AI. In: Nexus of AI, climatology, and urbanism for smart cities. IGI Global Scientific Publishing, pp 59–90
76. Skaržauskienė A, Mačiulienė M (2020, October) Mapping international civic technologies platforms. In: Informatics, Vol 7, No 4, p 46. MDPI
77. Shahidehpour M, Li Z, Ganji M (2018) Smart cities for a sustainable urbanization: Illuminating the need for establishing smart urban infrastructures. IEEE Electrification magazine 6(2):16–33
78. Shekhar S, Colletti J, Muñoz-Arriola F, Ramaswamy L, Krintz, C, Varshney, L, Richardson D (2017) Intelligent infrastructure for smart agriculture: an integrated food, energy and water system. arXiv preprint arXiv:1705.01993
79. Shukla RP (2025) Blockchain for enhanced transparency and traceability in circular supply chains. In: Innovating sustainability through digital circular economy. IGI Global Scientific Publishing, pp 203–228

80. Singh H, Singh BP (2025) Bridging e-governance and sustainability: unveiling synergies for a sustainable future. In: Leveraging futuristic machine learning and next-generational security for e-Governance. IGI Global Scientific Publishing, pp 1–24
81. Stecuła K, Wolniak R, Grebski WW (2023) AI-driven urban energy solutions—from individuals to society: a review. Energies 16(24):7988
82. Turskis Z, Šniokienė V (2024) IoT-driven transformation of circular economy efficiency: an overview. Mathematical and Computational Applications 29(4):49
83. Usman A, Ali AY (2024) The impacts of ICT gadgets on climate change and mitigative solutions. Journal of Computer, Software, and Program 1(2):1–7
84. Van Hoang T (2024) Impact of integrated artificial intelligence and internet of things technologies on smart city transformation. J Tech Educ Sci 19(Special Issue 01):64–73
85. Wongthongtham P, Marrable D, Abu-Salih B, Liu X, Morrison G (2021) Blockchain-enabled Peer-to-Peer energy trading. Comput Electr Eng 94:107299
86. Yuan B, Li J (2019) The policy effect of the General Data Protection Regulation (GDPR) on the digital public health sector in the European Union: an empirical investigation. Int J Environ Res Public Health 16(6):1070
87. Zhu Q, Duan Y, Sarkis J (2024) Supply chain carbon transparency to consumers via blockchain: does the truth hurt? The International Journal of Logistics Management 35(3):833–864

Chapter 3
China's Smart City Evolution: Leveraging ICTs for Sustainable Urban Growth

3.1 China's Smart City Vision and Sustainability Goals

Over the past few decades, China's rapid urbanization has changed the socioeconomic landscape of the country, but it has also brought about serious problems with resource consumption, environmental degradation, and strain on the infrastructure. Since more than 60% of the nation's population now lives in cities, there is growing pressure on the nation to maintain liveable, resilient, and efficient cities [20, 21]. China has adopted the idea of smart cities in response to these intricate urban issues [9], utilizing cutting-edge digital technologies to improve urban efficiency, sustainability, and governance [19]. The country's vision for smart cities, which aims to strike a balance between technological innovation, environmental stewardship, and social inclusivity, is closely linked to its larger sustainable development goals [26].

Rapid industrial growth, mass migration to cities, and extensive infrastructure expansion have all been hallmarks of China's urbanization. Although this has fuelled economic growth, it has also resulted in increased energy demands, air pollution, traffic jams, and environmental issues. In many cities, disparities in access to services are made worse by deteriorating infrastructure, ineffective resource management, and the growing urban–rural divide [10, 40]. By incorporating information and communication technologies (ICTs) to build more resilient and sustainable urban environments [13], China has made smart city development the focal point of its urban strategy in order to address these problems. This change is in line with the nation's larger objective of becoming carbon neutral by 2060, which calls for a significant overhaul of urban operations.

To promote smart city initiatives, the Chinese government has unveiled a number of important national strategies, with a focus on utilizing digital infrastructure, artificial intelligence, big data, and the Internet of Things (IoT). The New Urbanization Plan is a fundamental policy framework that emphasizes inclusive and sustainable urban growth [14]. This strategy, which was first introduced in 2014 and has

A. Cheshmehzangi, *Urban Innovation and Sustainability*, Urban Sustainability,
https://doi.org/10.1007/978-981-96-6370-5_3

since been updated, aims to improve resource efficiency, encourage green development, and improve urban planning. By promoting the growth of medium-sized cities, improving public services, and lessening the environmental impact of urban expansion, the plan places a strong emphasis on a people-centered approach [51]. The goal of the New Urbanization Plan is to create more liveable urban areas that strike a balance between environmental responsibility and economic growth [20] by incorporating smart city technologies.

Simultaneously, the Digital China initiative has been instrumental in hastening the implementation of smart city technologies [38]. Launched in 2017, this national strategy encourages the broad use of digital platforms to modernize public services, governance, and industries. Digital China has established the technological underpinnings for the success of smart cities by promoting advancements in cloud computing, blockchain, and 5G connectivity. The program encourages the creation of intelligent infrastructure, such as digital governance platforms, energy-efficient urban services, and intelligent transportation systems. Cities like Shanghai, Hangzhou, and Shenzhen have become prime examples of the effects of Digital China, demonstrating how ICTs can improve urban functionality while lowering environmental costs.

China's dedication to sustainability is closely linked to its smart city agenda. As part of its dual carbon strategy, which aims to peak carbon emissions by 2030 and achieve carbon neutrality by 2060, the government has given green and low-carbon urban development top priority in recognition of the environmental costs associated with rapid urbanization [37]. Using data-driven solutions to optimize energy consumption, lower emissions, and advance circular economies, smart cities are essential to this shift [48]. For example, smart grids and AI-driven energy management systems are being used to integrate renewable energy sources into urban power networks and increase the efficiency of electricity distribution [55]. In a similar vein, digital technologies are assisting cities in better managing waste, monitoring air quality, and optimizing water usage, guaranteeing that sustainability continues to be a fundamental aspect of urban growth.

China's widespread adoption of IoT and artificial intelligence has transformed urban management by making it possible to collect data in real-time and use predictive analytics to make better decisions [49]. For instance, AI-driven traffic monitoring is used by smart transportation systems to ease traffic and cut down on fuel usage. By examining traffic patterns and dynamically modifying signals to reduce congestion, Hangzhou's City Brain project—a historic smart city initiative—has effectively illustrated how artificial intelligence (AI) can enhance urban mobility [42]. In addition to improving commuter experiences, this also helps to improve air quality and reduce emissions. In a similar vein, developments in digital governance have simplified administrative procedures, increasing the effectiveness and accessibility of public services [15]. Through integrated digital platforms, citizens can now access legal, social security, and healthcare services, improving urban inclusivity and cutting down on bureaucratic inefficiencies.

China's dedication to smart city clusters and regional integration is another aspect of its smart city vision that sets it apart [46]. The government has created extensive projects like the Greater Bay Area Smart City Cluster, which connects cities like

Hong Kong, Shenzhen, and Guangzhou via integrated digital and transportation infrastructure, in recognition of the interdependence of urban economies [32, 41, 43]. By facilitating resource sharing, economic cooperation, and coordinated environmental policies, these regional smart city networks make sure that technological advancements benefit entire metropolitan regions rather than discrete urban centers.

Even with these bold plans, there are still obstacles to overcome before smart city technologies can be implemented in a way that is safe, effective, and equitable. The digital divide [23–25], cybersecurity threats, and data privacy issues present major barriers to the inclusive adoption of smart city projects [16]. A strong regulatory framework that strikes a balance between technological innovation, morality, and citizen rights is necessary to address these problems [7]. By enacting laws like the Personal Information Protection Law (PIPL), the Chinese government has improved data security [8, 27, 60] and made sure smart city solutions follow security and privacy guidelines.

With new technologies like blockchain, edge computing, and AI-driven automation playing bigger roles, China's smart city development is expected to continue on its current trajectory of rapid advancement [3, 60]. It will be essential to strike a balance between sustainability and technological advancement as cities grow more networked and dependent on digital systems (2 [1, 2, 17]. China hopes to establish itself as the world leader in smart city innovation by incorporating ICTs into sustainable urban planning, tackling the pressing issues of climate change, urban resilience, and equitable development [33]. In addition to influencing Chinese cities' futures, the realization of this vision will provide important insights for international urban sustainability initiatives.

3.2 ICT-Driven Innovations in Smart Cities in China

3.2.1 China's Progress to Date: AI, IoT, and Big Data for Urban Efficiency

China has revolutionized urban efficiency on a never-before-seen scale by incorporating big data, the Internet of Things (IoT), and artificial intelligence (AI) into its smart city infrastructure. These technologies are propelling advancements in smart grids, AI-powered urban planning, and intelligent transportation systems, streamlining city operations to boost sustainability and quality of life [28, 17]. Hangzhou's City Brain, an AI-driven system that has transformed urban mobility through real-time traffic optimization, is among the most well-known and effective examples of China's progress in smart cities. It exemplifies the revolutionary impact of digital solutions in urban management.

China's smart city initiatives, which aim to address the growing problems of emissions, traffic, and ineffective mobility networks, are centered on intelligent transportation systems [12]. AI-powered traffic monitoring has been implemented in cities like

Shenzhen and Shanghai, using real-time data gathered from thousands of sensors, security cameras, and GPS-enabled automobiles. AI systems can forecast traffic patterns, dynamically modify traffic signals, and maximize road usage by examining this enormous dataset [6]. This data-driven strategy has greatly cut fuel consumption and travel times, which is in line with China's larger sustainability objectives. Meanwhile, by utilizing IoT-enabled systems that continuously track electricity demand, smart grids are revolutionizing energy management. These grids improve overall energy efficiency, lower transmission losses, and make it easier to integrate renewable energy sources. For example, smart grid technology in Beijing has made it possible to seamlessly integrate wind and solar energy into urban electricity networks [44], lowering carbon emissions and reducing reliance on fossil fuels.

Hangzhou's City Brain, an AI-powered urban management system created by Alibaba Cloud in partnership with local authorities, is a notable example of China's progress toward smart cities. City Brain was first introduced in 2016 as a traffic management solution, but it has since spread to other urban sectors, showcasing the ability of AI and big data to increase citywide efficiency [34]. In order to optimize traffic flow, the system gathers enormous volumes of traffic data from video feeds, GPS units, and Internet of Things sensors. It then analyzes patterns in real time. Residents' commute times have been greatly shortened by City Brain's 15%–20% congestion reduction through dynamic traffic light adjustments and optimized route recommendations.

Beyond reducing traffic, City Brain offers additional benefits. By giving ambulance and firetruck routes priority, its AI-driven insights have been utilized to expedite emergency response times and guarantee that first responders can move through the city more effectively. To improve urban mobility, the system has also been integrated with public transportation networks, allowing bus and subway schedules to be synchronized with current traffic conditions. Over 20 Chinese cities, including Beijing and Shanghai, have adopted Hangzhou's City Brain because of its success [20], demonstrating its scalability and efficacy in urban management.

The results of the City Brain case study in Hangzhou demonstrate the enormous potential of big data, IoT, and AI to transform urban efficiency. Cities can improve energy distribution, ease traffic, and improve public services by utilizing automation and predictive analytics. Setting the standard for global urban innovation [22], China's advancements in smart city development show how digital technologies, when combined with urban infrastructure, can produce more liveable, efficient, and sustainable cities.

3.2.2 *Digital Governance and Citizen-Centric Services in China*

China has made great progress in putting in place digital governance systems that use ICT advancements to improve citizen engagement and expedite government operations. Cities are now able to offer their citizens more effective, transparent, and easily accessible services thanks to the integration of e-governance, smart healthcare, and digital public services [18]. These programs support the overarching objectives of developing smart cities that are both technologically sophisticated and receptive to the needs of their citizens. Shenzhen's digital governance model is one of the most well-known instances of digital governance in China [50]. It has established a standard for how digital technologies can improve civic engagement and change the way government services are delivered.

In China, e-governance has been developing quickly, with an emphasis on improving the efficiency, transparency, and accessibility of government services [29, 57]. Chinese cities are shifting toward more efficient, user-friendly platforms that minimize bureaucratic inefficiencies and guarantee quicker service delivery by digitizing public services and administrative procedures. With the launch of the Shenzhen e-Government Service Platform, Shenzhen has taken the lead in this shift [53]. Residents can use this platform to access a variety of government services through a single online portal, such as public welfare, social security, and business registration. Through the use of big data and cloud computing, Shenzhen's digital governance model makes government interactions more efficient and less time-consuming for citizens by reducing the need for in-person visits, waiting times, and paperwork.

Another area where China's digital governance has advanced significantly is smart healthcare. Digital technologies are being used in cities like Shenzhen to lower costs, increase access to medical services, and enhance the general standard of care. Doctors can more easily access medical records and deliver individualized care thanks to the Shenzhen Health Information Platform, which compiles patient data from clinics, hospitals, and health insurance companies. Telemedicine services are also available to residents, increasing access to healthcare, especially in underprivileged areas. These developments have increased the effectiveness of the healthcare system and improved public health outcomes, proving the ability of ICT to improve necessary services.

The emphasis on citizen-centric services, which give residents' needs and convenience top priority, is one of the most noteworthy features of Shenzhen's digital governance model. Personalized services are made possible by the city's use of AI, big data, and IoT, where government systems proactively recognize and respond to people's needs [18]. For instance, Shenzhen can automatically produce solutions that enhance the quality of life for locals by using AI-driven data analysis to monitor urban issues like waste management, pollution, and traffic congestion in real time [5]. Rapid responses to problems like waste disposal, emergency services, and environmental monitoring are made possible by the integration of digital platforms with urban services, which increases the city's adaptability and responsiveness.

Because of Shenzhen's success, other Chinese cities have looked to it as a model for how digital governance can create more inclusive, transparent, and effective urban environments [18]. In order to increase both operational effectiveness and citizen satisfaction, the city's strategy places a strong emphasis on the smooth integration of technology into public services [45, 59]. Shenzhen is extending its e-governance platforms' capabilities to encompass additional facets of urban life, such as social welfare, education, and public safety, as they continue to evolve [39]. In addition to providing a global model for how ICT can propel sustainable urban development, this progressive approach to digital governance is advancing the larger objectives of building smart cities that improve the standard of living for all citizens. The results of Shenzhen's digital governance model highlight the enormous potential of ICT to change the dynamic between citizens and governments and build more effective, responsive, and equitable urban environments.

3.3 China and Facing the Challenges with Strategic Responses

3.3.1 Managing Data Privacy, Security, and Ethical Concerns

The integration of digital technologies like artificial intelligence (AI), the Internet of Things (IoT), and big data raises serious concerns about data privacy, cybersecurity, and the ethical use of information as China develops its smart cities [18]. As the nation expands its urban management systems, which mainly depend on real-time data collection and analysis, these issues become even more urgent. The success of China's smart city projects depends on how well these problems are managed, all the while protecting citizens' rights and preserving faith in governmental processes [21]. Although China's changing data protection laws and regulatory frameworks are crucial in resolving these issues, issues with public trust and cybersecurity risks still exist.

The enormous volume of private and sensitive data produced by digital services is one of the main issues with China's smart city development. Although the collection and use of this data raise privacy concerns, it is crucial for effective urban management. In response, the Chinese government has created a number of regulatory frameworks to strengthen data security and guarantee the moral use of private data [18]. A historic law that established standards for the gathering, storing, and processing of personal data is the Personal Information Protection Law (PIPL), which went into force in 2021. The PIPL seeks to provide precise guidelines for consent, data access, and security for both the public and private sectors. It is based on the General Data Protection Regulation (GDPR) of the European Union. It strengthens citizens' rights to manage their personal information and requires more stringent oversight of data handling.

Cybersecurity threats still pose a serious problem in spite of these developments. The interconnected systems that underpin smart cities are susceptible to cyberattacks, which could jeopardize the security of the data and the infrastructure they support. China has responded by tightening its cybersecurity regulations, such as the Cybersecurity Law (passed in 2017), which requires the protection of data stored inside China's borders and the implementation of security measures for vital infrastructure [18]. The purpose of this law is to guard against data breaches, hacking, and illegal access to private data. Public trust is still at risk from cybersecurity threats and breaches, though, particularly in light of the government's extensive data collection on citizens' activities.

China's continuous creation of an extensive data protection framework serves as a crucial case study demonstrating its strategic responses to data privacy and security issues. These initiatives seek to create a safe atmosphere for the development of smart cities in addition to enhancing public trust. To maintain transparency and protect citizen information, data privacy measures are incorporated into the architecture of urban management systems in places like Beijing and Hangzhou [54]. It is still difficult to strike a balance between encouraging innovation and guaranteeing strong data protection, though. Although China's government is committed to addressing privacy and security concerns, as evidenced by its evolving cybersecurity strategies and data protection laws, it is evident that ongoing adaptation will be necessary to address new challenges in an increasingly digital urban environment.

3.3.2 Balancing Technological Growth with Environmental Sustainability

One of the biggest challenges facing China as it speeds up its smart city initiatives is striking a balance between the need for environmental sustainability and the quick advancement of technology. ICT advancements are the driving force behind the nation's ambitious urban development goals, but they must be combined with energy-saving and environmentally friendly measures. The development of low-carbon cities, energy-efficient urban infrastructure, and green ICT initiatives are essential to tackling this issue. Utilizing digital technologies to optimize resource use, lower carbon emissions, and develop sustainable urban environments is the main goal of these initiatives. But maintaining this equilibrium calls for meticulous preparation and a dedication to sustainable development methods [30].

Urban systems' environmental impact is being lessened in large part by green ICT initiatives. These projects entail the development and implementation of energy-efficient technologies that reduce the energy usage of digital infrastructure, including sensors, data centers, and communication networks. By encouraging the use of renewable energy sources, energy-saving technologies, and environmentally friendly materials in ICT systems, the Chinese government has incorporated green ICT principles into its plans for smart cities. One such initiative that improves energy efficiency and facilitates the switch to renewable energy is smart grids, which combine digital technologies to optimize energy distribution.

A crucial element of China's smart city vision is energy-efficient urban infrastructure. Real-time energy consumption optimization is made possible by the incorporation of IoT, AI, and big data into urban systems [18]. Energy management systems installed in smart buildings can automatically modify heating, cooling, and lighting in response to weather and occupancy trends. This promotes sustainability objectives and lessens energy waste. In addition, Chinese cities are implementing energy-efficient design guidelines and green building certifications with the goal of lowering energy consumption in both the public and private sectors.

Beijing's clever energy management techniques are among the best examples of China's attempts to strike a balance between environmental sustainability and technological advancement. In order to optimize its energy consumption across multiple sectors, the city has implemented cutting-edge digital technologies. The implementation of smart grids, which employ AI and big data to continuously monitor energy consumption and make dynamic adjustments, is one noteworthy initiative. By incorporating renewable energy sources like wind and solar into the city's electrical grid, these smart grids lessen reliance on fossil fuels and cut carbon emissions. Beijing has also incorporated energy-efficient technologies into its transportation networks, such as AI-powered traffic management systems that optimize vehicle flow and lower fuel consumption, and electric vehicle (EV) charging stations.

The city's environmental impact has decreased as a result of these clever energy management techniques. In addition to increasing energy efficiency, Beijing's initiatives have served as a model for other Chinese and international cities, illustrating how ICT advancements can help build sustainable, low-carbon urban environments. There are still obstacles to overcome, though, especially in terms of guaranteeing the initiatives' scalability and getting past technological obstacles to incorporate renewable energy sources throughout the city. Beijing's clever energy management techniques, however, offer a striking illustration of China's efforts to strike a balance between environmental sustainability and technological advancement.

3.4 China's Future Trajectories and Global Implications

As China continues to push the boundaries of smart city development, the country's future trajectories in urban innovation hold significant implications not only for its own cities but also for the global urban sustainability landscape. The incorporation of cutting-edge technologies like 5G, blockchain, and artificial intelligence (AI), which have the potential to completely alter how urban areas operate, interact, and change, is essential to China's next-generation smart cities [18]. By improving connectivity, data transparency, and decision-making, these technologies have the potential to build smarter [21], more resilient cities that can withstand the challenges of resource scarcity, rapid urbanization, and climate change.

China's ongoing projects are already showing how 5G, blockchain, and AI will shape the nation's future smart cities. The foundation for the interconnection of urban infrastructures will be 5G technology, which offers low latency and fast data transfer. Cities will be able to better manage everything from waste management to traffic flow thanks to 5G's ability to enable real-time communication between devices, sensors, and systems [18]. Contrarily, blockchain will offer a transparent and safe platform for handling identities, transactions, and data, boosting confidence in digital governance and guaranteeing the quality of urban services. Blockchain has the potential to significantly increase the security and transparency of urban systems by generating tamper-proof records and streamlining procedures like energy trading and supply chain management. Artificial intelligence (AI) technologies, which are already being used in urban planning and traffic optimization, will be crucial in analyzing large volumes of data in order to forecast trends, allocate resources optimally, and improve decision-making. When combined, these technologies will produce hyperconnected, data-driven cities with more individualized services and responsive governance.

There are important worldwide ramifications as China moves closer to these lofty objectives. For other countries aiming to achieve urban sustainability, China's developing smart city ecosystem provides valuable insights [35]. The Chinese model highlights the difficulties in managing data privacy, equity, and sustainability while also showcasing the transformative potential of extensive technological integration and government-led innovation. The significance of coordinating technology developments with sustainability goals is one important lesson to be learned from China's smart city development [18]. In order to guarantee that the technologies implemented not only maximize efficiency but also support sustainable urban ecosystems, the drive for smart cities must be based on a vision of social justice and environmental stewardship. Furthermore, China's focus on digital governance—which is demonstrated by initiatives like Shenzhen's e-Government Service Platform [11]—offers a useful perspective on how to preserve and strengthen public trust via openness, accessibility, and the application of data to better the lives of citizens.

China's smart city experience offers a priceless model for accelerating digital transformation while tackling urban challenges for global urban sustainability initiatives. The difficulty, though, is in reproducing these achievements in other settings where variables like political structures, legal frameworks, and the degree of technological infrastructure may vary. However, China's emphasis on innovation and scalability is likely to have an impact on global trends, inspiring other cities to integrate smart technologies into their plans for sustainable urban growth [18]. Looking ahead, China's smart city trajectory will continue to influence global urban development, providing lessons and inspiration for future cities on how to combine sustainability and technology.

3.5 The Story of Hainan and Its Significance for ICT-Driven Innovation and Sustainability Pathways

As we wrap up this chapter, it is important to consider Hainan's (the book's case study example) transformative journey and its importance in forming ICT-driven innovation and sustainability pathways. Hainan's methodical approach to fusing sustainability and technology provides insightful information about how regional development can be both forward-thinking and ecologically responsible [52]. A notable precedent for how technology can be used to create more resilient, sustainable, and inclusive urban environments has been established by the island's efforts to use ICT for urban development. The four main facets of sustainability—social, environmental [31], institutional, and economic—are all balanced in Hainan's dedication to sustainability. The island's development trajectory is greatly influenced by each of these pillars, which make sure that long-term sustainability is not sacrificed in the name of technological advancement.

From the standpoint of social sustainability, Hainan's ICT-driven projects have concentrated on raising the standard of living for its citizens. Through the use of digital technologies like smart education systems, digital healthcare, and e-governance, Hainan has improved public services and encouraged increased community involvement. A more inclusive society is produced when ICT is incorporated into public services to guarantee that people have improved access to social welfare, healthcare, and education [18]. Furthermore, Hainan's focus on digital platforms for public participation has improved civic engagement and given communities the ability to participate in urban planning and decision-making processes [56, 58]. This approach to social sustainability emphasizes how technology can be used to increase social equity and inclusion in addition to being an efficient tool.

Another important aspect that Hainan has addressed with its ICT initiatives is environmental sustainability. The island's attempts to incorporate smart grids, renewable energy, and smart agriculture into its development model serve as an example of how technology can help protect the environment. Hainan has been able to minimize waste, lessen its environmental impact, and encourage sustainable practices in energy, transportation, and agriculture by using digital technologies to optimize resource management [36]. Incorporating IoT sensors and artificial intelligence (AI) into resource management systems guarantees more effective use of energy, water, and other essential resources, supporting Hainan's overall environmental objectives [4]. These programs act as a reminder that ecological preservation and technological advancement can and should coexist, creating a sustainable basis for coming generations.

Economically speaking, Hainan's use of ICT has stimulated new growth prospects and created an atmosphere that is favorable to entrepreneurship, innovation, and economic diversification. The development of digital economies has been aided by the creation of free trade zones and smart industry hubs, which have increased employment and supported regional industries. Hainan is establishing itself as a model for economic sustainability in the digital age by emphasizing high-tech industries like digital agriculture, tourism, and health services. Long-term economic resilience is also aided by the island's focus on green innovation, such as sustainable agriculture and clean energy technologies. These innovative business practices guarantee Hainan's competitiveness in a global economy that is becoming more digitally and ecologically aware.

Lastly, the role of governance, policy-making, and regulatory frameworks that direct Hainan's smart city initiatives demonstrates the institutional sustainability of the city's development. By enacting laws that promote innovation while guaranteeing that social, environmental, and economic considerations are given top priority, the government of Hainan has been instrumental in laying the groundwork for ICT-driven sustainability [47]. The island's strategy for digital governance makes decision-making more effective while guaranteeing accountability, transparency, and the alignment of technological advancement with sustainable development objectives. The existing regulatory frameworks address important concerns like data privacy, cybersecurity, and fair access to technology while encouraging responsible ICT adoption. In order to keep Hainan's trajectory toward ICT-driven innovation in line with its long-term sustainability goals, this institutional framework is essential.

We conclude the chapter by considering our main conclusions, which are illustrated in the book's other chapters. We think Hainan is a compelling example of how ICT can be used to propel sustainable urban development in all its dimensions, based on our (nearly) two years of work, fieldwork, experimental studies, and pilot projects centered on ICT-driven approaches in achieving resilience, smartness, and sustainability at various scales. Hainan is proving that it is feasible to create future cities that are not only technologically sophisticated but also socially inclusive, environmentally conscious, economically resilient, and institutionally sound by fusing technology with a holistic sustainability framework. The island's strategy provides important insights on how to strike a balance between innovation and sustainability, guaranteeing that the advantages of ICT-driven growth are felt by everyone, as other nations and regions seek to duplicate Hainan's success, possibly through experimental studies of varying scales. The experience of Hainan serves as evidence of how smart cities can serve as role models for sustainable development in the era of digitalization.

References

1. Appio FP, Lima M, Paroutis S (2019) Understanding smart cities: innovation ecosystems, technological advancements, and societal challenges. Technol Forecast Soc Chang 142:1–14
2. Aurigi A (2006) New technologies, same dilemmas: policy and design issues for the augmented city. J Urban Technol 13(3):5–28
3. Badidi E (2022) Edge AI and blockchain for smart sustainable cities: promise and potential. Sustainability 14(13):7609
4. Bi L (2011) Integrating planning theory with energy planning in developing rural areas: a critical assessment of the energy intervention programs in rural Hainan, China. Available online from: https://uwspace.uwaterloo.ca/handle/10012/5805
5. Bosco G, Riccardi V, Sciarrone A, D'Amore R, Visvizi, A (2024) AI-driven innovation in smart city governance: achieving human-centric and sustainable outcomes. Transforming government: people, process and policy. Available online from: https://www.emerald.com/insight/content/doi/10.1108/TG-04-2024-0096/full/html
6. Boukerche A, Tao Y, Sun P (2020) Artificial intelligence-based vehicular traffic flow prediction methods for supporting intelligent transportation systems. Comput Netw 182:107484
7. Brownsword R, Yeung K (eds) (2008) Regulating technologies: legal futures, regulatory frames and technological fixes. Bloomsbury Publishing, Oxford
8. Calzada I (2022) Citizens' data privacy in china: The state of the art of the personal information protection law (pipl). Smart Cities 5(3):1129–1150

9. Caprotti F, Cowley R, Bailey I, Joss S, Sengers F, Raven R, Spaeth P, Jolivet E, Tan-Mullins M, Cheshmehzangi A, Xie L (2017) Smart eco-city development in Europe and China: policy directions. Exeter: University of Exeter, Available from: https://eprints.gla.ac.uk/182634/
10. Chen C, LeGates R, Zhao M, Fang C (2018) The changing rural-urban divide in China's megacities. Cities 81:81–90
11. Chen H, Su Z (2024) Role of smart city for sustainable development: exploring the nexus among smart cities, e-governance and environmental development. Int J Electron Bus 19(3):255–270
12. Chen Y, Ardila-Gomez A, Frame G (2017) Achieving energy savings by intelligent transportation systems investments in the context of smart cities. Transp Res Part D: Transp Environ 54:381–396
13. Cheshmehzangi A (2016) City enhancement beyond the notion of "Sustainable City": Introduction to integrated assessment for city enhancement (iACE) toolkit. Energy Procedia 104:153–158
14. Cheshmehzangi A (2016) China's New-Type Urbanisation Plan (NUP) and the foreseeing challenges for decarbonization of cities: a review. Energy Procedia 104:146–152
15. Cheshmehzangi A (2022) The impacts of ICT platforms on public place Use in China. Built Environment 48(2):264–279
16. Cheshmehzangi A (2022b) The role of ICT and smartness for peace-building processes and practices. In: ICT, cities, and reaching positive peace. Singapore: Springer Nature Singapore, pp 17–46
17. Cheshmehzangi A (2022) ICT, cities, and reaching positive peace. Springer, Singapore
18. Cheshmehzangi A, Dawodu A, Sharifi A (2021) Sustainable Urbanism in China. Routledge, New York and London
19. Cheshmehzangi A, Pourroostaei Ardakani S (2024) Navigating the future with the nexus between digital twin computing and urban intelligence. In: Pourroostaei Ardakani S Cheshmehzangi A (eds). Digital twin computing for urban intelligence. Singapore: Springer Nature Singapore, pp 245–256
20. Cheshmehzangi A, Tang T (2024) 30 years of urban change in China's 10 core cities. Springer, Singapore
21. Cheshmehzangi A, Tang T (2024) China under construction: shaping cities through recent urban transformation. Springer, Singapore
22. Cheshmehzangi A, Zou T (2024) Literary analysis of the connection between Urban Innovations and Sustainable Development Goals (SDGs). Sustainability 16(15):6405
23. Cheshmehzangi A, Zou T, Su Z (2022) The digital divide impacts on mental health during the COVID-19 pandemic. Brain Behav Immun 101:211–213
24. Cheshmehzangi A, Zou T, Su Z, Tang T (2023) The growing digital divide in education among primary and secondary children during the COVID-19 pandemic: An overview of social exclusion and education equality issues. J Hum Behav Soc Environ 33(3):434–449
25. Cheshmehzangi A, Zou T, Su Z, Dawodu A, Ardakani SP, Sedrez M, Tang T (2024) Growing digital divide for expatriates population in China: a user perspective analysis during the COVID-19 Pandemic. SAGE Open 14(4):21582440241274612
26. China Academy of Information, Communi, EU-China Policy Dialogues Support Facili (2015) Comparative study of smart cities in Europe and China 2014. Singapore and Germany: Springer
27. Creemers R (2022) China's emerging data protection framework. J Cybersecur 8(1):tyac011
28. Djahel S, Doolan R, Muntean GM, Murphy J (2014) A communications-oriented perspective on traffic management systems for smart cities: Challenges and innovative approaches. IEEE Communications Surveys & Tutorials 17(1):125–151
29. Du P, Yu S, Yang D (2019) The development of e-governance in China. In: Research series on the Chinese dream and China's development path. Springer Singapore
30. Faroqi H, Irajifar L, Cheshmehzangi A (2022) Sustainable development in smart and resilient local government areas: an empirical investigation of Victoria, Australia. Int J Sustain Dev & Plan 17(6)
31. Feng H, Liu M, Xu M, Zhang M, Mo L, Chen T, Tan X, Liu Z (2021) Study on the integrated protection strategy of water environment protection: the case of Hainan Province of China. Environ Technol Innov 24:101990

32. Govada SS, Lau H, Kotala S (2022) Towards a smart megacity: Hong Kong and Shenzhen. Smart global megacities: Collaborative research: Tokyo, Mumbai, New York, Hong Kong-Shenzhen, Kolkata, 317–380
33. Hong CY, Fan HY (2019) Public coordination and institution innovation: an institutional approach for hub-and-spokes logistics network in the Greater Bay Area. Journal of Public Policy and Administration 3(2):39
34. Janus D (2021) Smart cities in China: sustainable or surveyed. Sprawy Międzynarodowe 74(1):153–174
35. Kuhn BM (2018) China's commitment to the sustainable development goals: an analysis of push and pull factors and implementation challenges. Chinese Political Science Review 3(4):359–388
36. Li C, Zhang, Z (2025) Collaborative development of the digital economy, tourism economy, and ecological environment to achieve green and sustainable development. Environ, Dev Sustain, 1–23
37. Liao K, Liu J (2024) Digital infrastructure empowerment and urban carbon emissions: Evidence from China. Telecommunications Policy 48(6):102764
38. Liu B, Li Y, Liu J, Hou Y (2024) Does urban innovation policy accelerate the digital transformation of enterprises? Evidence based on the innovative City pilot policy. China Econ Rev 85:102167
39. Liu T, Yang X, Zheng Y (2020) Understanding the evolution of public–private partnerships in Chinese e-government: four stages of development. Asia Pacific Journal of Public Administration 42(4):222–247
40. Long H, Tu S, Ge D, Li T, Liu Y (2016) The allocation and management of critical resources in rural China under restructuring: problems and prospects. J Rural Stud 47:392–412
41. Ma S, Ding J, Huang Z, Guo R (2024) Evaluation of the urban industrial coupling strategy based on the global production networks theory: a case study of the smart phone industry in the Guangdong-Hong Kong-Macao Greater Bay Area. PLoS ONE 19(4):e0300588
42. Noesselt N (2020) City brains and smart urbanization: regulating 'sharing economy' innovation in China. Journal of Chinese Governance 5(4):546–567
43. Ogawa A, Tsuchiya T (2024) China's smart port initiative in the Guangdong–Hong Kong–Macao Greater Bay Area. Available online from: https://mpra.ub.uni-muenchen.de/121687/
44. Rehmani MH, Reisslein M, Rachedi A, Erol-Kantarci M, Radenkovic M (2018) Integrating renewable energy resources into the smart grid: Recent developments in information and communication technologies. IEEE Trans Industr Inf 14(7):2814–2825
45. Ren Q, Zhou X, Yang X (2024) Comparison of intelligent community design from the perspective of Modernization-Shenzhen and Singapore as examples. Urban Geospatial Planning and Sustainability 1(1):1–24
46. Riva Sanseverino E, Riva Sanseverino R, Anello E (2018) A cross-reading approach to smart city: a european perspective of chinese smart cities. Smart Cities 1(1):26–52
47. Shou M, Jia F, Yu J, Wu Y (2025) Challenges in developing information and communication technology (ICT) use for rural e-governance: an ecology perspective. Inf Syst J 35(1):247–278
48. Song T, Dian J, Chen H (2023) Can smart city construction improve carbon productivity?—a quasi-natural experiment based on China's smart city pilot. Sustain Cities Soc 92:104478
49. Wei Y, Yuan H, Li H (2024) Exploring the contribution of advanced systems in smart city development for the regeneration of urban industrial heritage. Buildings 14(3):583
50. Wong W, Chu M (2020) Digital governance as institutional adaptation and development. China Review 20(3):43–70
51. Wu Y (2025) Improving systems and mechanisms for people-centred urbanization. In: A study on people-centered urbanization. Singapore: Springer Nature Singapore, pp 297–351
52. Xiang H, Farid HMA, Riaz M (2024) Linear programming-based fuzzy alternative ranking order method accounting for two-step normalization for comprehensive evaluation of digital economy development in provincial regions. Axioms 13(2):109
53. Xie S, Luo N, Yarime M (2024) Data governance for smart cities in China: the case of Shenzhen. Policy Design and Practice 7(1):66–86

54. Yan X, Li T (2023) Construction and application of urban digital infrastructure—practice of "Urban Brain" in facing COVID-19 in Hangzhou, China. Eng Constr Archit Manag 30(8):3123–3141
55. Yu X, Xue Y (2016) Smart grids: a cyber–physical systems perspective. Proc IEEE 104(5):1058–1070
56. Zhang J, Chen S, Cheng C, Liu Y, Jennerjahn TC (2023) Citizen science to support coastal research and management: Insights from a seagrass monitoring case study in Hainan China. Ocean & Coastal Management 231:106403
57. Zhang J (2006) Good governance through e-governance? assessing China's e-government strategy. Journal of E-government 2(4):39–71
58. Zhang J, Fedder B, Wang D, Jennerjahn TC (2022) A knowledge exchange framework to connect research, policy, and practice, developed through the example of the Chinese island of Hainan. Environ Sci Policy 136:530–541
59. Zhang L (2019) Traffic congestion reduction and smart city strategy-a case study in Shenzhen, China. Master's thesis, University of Twente
60. Zhou MM, Qu Z, Wan J, Wen B, Yao Y, Lu Z (2024) Understanding Chinese internet users' perceptions of, and online platforms' compliance with, the Personal Information Protection Law (PIPL). Proceedings of the ACM on Human-Computer Interaction 8(CSCW1):1–26

Chapter 4
Innovation and Environmental Sustainability: ICT and Beach Management

4.1 An Overview of ICT Use for Beach Areas

Addressing coastal issues from a beach management standpoint is a key component of coastal areas' sustainability agenda. Environmental sustainability through the implementation of various beach management policies and practices has been the main focus of coastal zone management studies [12]. Such studies focus on primary aspects of sustainable tourism [18], preservation strategies and resource management [35], beach quality improvement for sustainable coastal tourism [22], and environmental protection [25]. The latter aspect has been a primary challenge in rapidly developing countries like China. Nonetheless, with the emergence of ICT-enabled technologies [10], beach management has been further improved from multiple perspectives of safety management [1]. cleanliness regarding beach pollution issues [33], and fisheries management [19]. For the latter, which is critical to coastal environmental protection, the concept of beach management units (BMUs) is developed to ensure fishery resources are better managed [26]. The objectives of such group formation are correlated to mainly fishery communities in coastal areas [28] and not necessarily for environmental protection purposes. Hence, there is still a major gap in ICT use for beach areas and coastal environmental protection strategies. This chapter aims to address this gap by opening up a new discourse for future research on ICT-enabled innovation for beach management. By exploring the potential applications of emerging technologies, this chapter seeks to provide a foundation for developing more effective, sustainable, and scalable solutions for coastal zone management. Ultimately, this research can pave the way for a deeper understanding of how ICT can contribute to preserving coastal ecosystems and promoting responsible tourism practices.

A recent study by [28] had a take on developing an ICT-based index for beach quality management. In their evaluation system, ICTs were introduced to manage and monitor the beach areas, which goes beyond the traditional participatory mechanisms for coastal management and their associated environmental protection strategies.

A. Cheshmehzangi, *Urban Innovation and Sustainability*, Urban Sustainability,
https://doi.org/10.1007/978-981-96-6370-5_4

Their study [28] focuses on using computerization and automation methods in beach management to ensure beach quality is managed and monitored through ICT-enabled tools and techniques. It is already widely acknowledged that the coastal ecosystem is primarily utilized by human activities of various sorts [16], from mass tourism [3] to uncontrolled fishing and pollution. This study mainly focuses on coastal fishing activities, using a case study example to expand on existing literature on fishing communities and beach management strategies that seem essential in touristic areas. The case of Sanya and five of its popular bays (i.e., famous for tourism mainly) are selected to provide a brief overview of ICT-enabled methods for beach management, particularly for the management and monitory of fishing and fishery. This is an area that requires further attention from the sustainable development perspective and in particular, is correlated with the development of sustainable cities and communities.

It is important to understand that ICT integration in coastal management can offer creative answers to the problems caused by pollution, overfishing, and pressures from tourism. Communities and stakeholders can collaborate to support sustainable fishing methods, lessen environmental degradation, and guarantee the long-term health of coastal ecosystems by utilizing real-time data, remote sensing technologies, and digital platforms. Furthermore, the development of adaptive management strategies that can adapt to the changing needs of local communities and the dynamic nature of coastal environments depends heavily on these ICT-enabled approaches.

While fishery trends and resource use have been critical arguments of beach management systems across the globe [2, 15, 20, 31], there are still many examples that suggest failures of such programs [34]. Such failures mainly occur due to the deployment of easy solutions [24], impacts of external forces and changes on traditional systems [30], the absence of long-term plans and visions [37], etc. More importantly, scholarly research is only limited to some case study examples and is more related to community-based development, actor-based design of beach management systems [21], and in line with general sustainable development ideas [11]. Thus, we see only a few ICT-related research on coastal and beach management areas, primarily focused on case study projects on coastal resource management [14], data sharing management schemes for fishery information [32], and benefit the aquaculture development projects [23]. Based on the existing literature and scholarly research, there is an urgent need for knowledge sharing and ICT-enabled innovation to ensure environmental protection is considered in the beach management system [7]. This approach is particularly effective in the face of growing illegal fishing and fishery activities in coastal areas. By using a case study example, this study opens up debates on the inclusion of ICT-enabled innovation for beach management systems and strategies.

Because of the ongoing environmental degradation that coastal regions experience as a result of both natural and man-made pressures, it is imperative that advanced technology be incorporated into their management in order to promote more sustainable practices [6]. Although earlier research has emphasized the value of traditional management techniques and local community involvement [5], there is a glaring lack of use of ICT solutions that can improve decision-making processes for stakeholders and local authorities alike and offer real-time data [29]. By investigating how ICT can be used to create more intelligent, effective, and scalable systems for managing coastal resources, stopping illicit activity, and fostering sustainable tourism, this study seeks to close that gap. By doing this, it hopes to support the creation of stronger and more resilient beach management plans, guaranteeing that environmental issues are sufficiently taken care of in coastal areas that are developing quickly.

4.2 Setting the Scene: Current Challenges and Facing the Complexity of Beach Management Issues in Sanya, Hainan Island

Conflicts over the use of coastal resources are becoming more frequent in Sanya, China's top coastal recreation destination, especially when it comes to striking a balance between the needs of environmental preservation, local livelihoods, and tourism. The quick growth of Sanya, a popular beach recreation area, has resulted in overlapping and frequently conflicting activities along its shorelines and nearshore waters. High-quality coastal environments are essential to the region's tourism-driven economy, but enduring problems like illicit fishing, environmental degradation, and inadequate monitoring jeopardize these resources' long-term sustainability. These issues have proven difficult for traditional beach management tools like CCTV surveillance and regulatory zoning to handle, so a more thorough, tech-driven approach to coastal governance is required.

Among the most pressing are the conflicts that still exist between recreational beachgoers and nearshore fishing operations despite existing regulations. In addition to depleting marine resources, illegal fishing practices negatively impact the overall experience of beachgoers, public safety, and aesthetic appeal. Nets and illicit fishing gear endanger swimmers, and overfishing creates ecological imbalances that devalue the region's marine environment. The issue is made worse by the fishing industry's political and economic significance, which has historically enjoyed strong local support and regularly opposes restrictive policies. Beach Management Units (BMUs) were created to coordinate fishing operations and encourage sustainable management, but they have not worked out well because of lax enforcement, a lack of integration with larger frameworks for coastal governance, and a failure to consider real-time data on user and environmental dynamics.

A rational and sustainable approach to resource management in Sanya requires thorough, current information on nearshore and shoreline activities. Understanding local fishery operations, trends in recreational use, and ecological health indicators is essential to make well-informed policy decisions that strike a balance between economic, social, and environmental priorities. The long-term viability of these coastal resources depends on improved data collection methods that can support proactive interventions and adaptive management. There is a revolutionary opportunity to enhance management responses through the use of ICT-enabled solutions, such as real-time monitoring networks, data-sharing platforms, and predictive analytics. ICT can support a dynamic, responsive beach management framework that guarantees the preservation of Sanya's priceless coastal assets while resolving the fundamental disputes between conflicting user groups by overcoming the limitations of static surveillance systems.

Sanya's beach management problems are particularly complex because of its dual status as a popular international tourist destination and a site of deeply embedded local economic practices. Compared to other Chinese coastal regions, Sanya's beaches receive millions of visitors annually from both domestic and foreign nations, putting more pressure on the country's natural resources and raising the stakes for maintaining environmental quality. This complexity is increased by the coexistence of large tourism infrastructure and small-scale, usually illegal fishing operations, which are less prevalent and politically entrenched in other coastal cities in China.

While Sanya's management is still dispersed among tourism authorities, local fishing collectives, and municipal governance, many popular beach destinations around the world, like those in the Mediterranean or Southeast Asia, rely on integrated coastal zone management (ICZM) frameworks that balance tourism and conservation. Sanya's strategy mainly depends on CCTV monitoring, which is unable to gather detailed information on both legal and illicit coastal activities, in contrast to situations in Australia or the US where sophisticated ICT frameworks are integrated into public–private management systems. Additionally, Sanya's Beach Management Units (BMUs) are largely concerned with controlling fishing operations without sufficiently addressing environmental protection or conflicts arising from public use [4], whereas other regions employ community-based conservation strategies. Sanya's situation is particularly difficult because of this disjointed approach and the political sensitivity of regulating traditional fishing methods. It also highlights the urgent need for creative ICT-enabled solutions to close the gap between resource equity, environmental sustainability, and enforcement.

4.2.1 Geographical Information About Sanya, Hainan Islands, China

Situated in the South China Sea, Sanya is the southernmost city of Hainan Island, which is the only tropical island province in China. Sanya, the island's second-biggest city after Haikou, the provincial capital, is vital to the area's tourism, trade, and maritime industries. The island itself connects China with Southeast Asia and beyond thanks to its advantageous location at the intersection of important shipping lanes and economic corridors in the Asia–Pacific area (Fig. 4.1).

Across the Qiongzhou Strait from Guangdong Province, Hainan Island is located just off the southern coast of mainland China. It is a crucial hub for global connectivity and trade because it is geographically closer to Vietnam and the Philippines than it is to northern China. With a sophisticated transportation system that includes Sanya Phoenix International Airport, which links the island to important cities in China,

Fig. 4.1 The Location of Sanya and Hainan Island and their location in the region of South China and South East Asia with Neighbouring countries such as Vietnam. *Source* royalty-free vector graphics and clip art from iStock on Sanya, Hainan Island, China

Southeast Asia, and other countries, Sanya in particular acts as a gateway for air and sea travel. Hainan's status as a regional hub for tourism and the economy is further enhanced by its participation in China's Belt and Road Initiative. Beyond its economic importance, Sanya is one of China's most popular tourist destinations due to its tropical climate and coastal setting. Because of its location along the South China Sea, which offers abundant marine biodiversity, it is a popular destination for ocean-related businesses like fishing and ecotourism. Sanya is positioned as a major player in regional development and coastal sustainability initiatives due to its advantageous climate and strategic location, which further emphasize the island's significance in China's larger economic, geopolitical, and environmental strategy.

4.3 Methods, Data Collection, and Analyses Strategy

The study is conducted in three steps: (1) Case study selection, (2) On-site mapping analysis, and (3) Interviews and the analysis of official reports, documents, and policies. First, case study selection is conducted based on three factors, i.e., the relevance of case study examples according to the context and the project objectives, availability of data for further analysis, and accessibility for on-site mapping analysis. Based on the contextual analysis, five bays in the City of Sanya, Hainan Island, are selected. All five cases are popular beach areas, with the presence of both tourists and local communities. The second step includes on-site mapping analysis, including three mapping studies. The first mapping study is to identify and confirm the level of population density and popularity in each of the selected five cases.

This mapping is conducted in three site visits to each case in three different periods of low season, mid-season, and high season for tourism activities. The research team conducted all site visits in September 2021 (low season), December 2021 (mid-season), and February 2022 (high season). These periods are selected to avoid any biased analysis of only one season and ensure data is consistent for all five case studies. This mapping is based on on-site daily records of population density in all five bays, without considering the main variables of rainy weather conditions and weekend periods. The mapping is done on-site by the research team members in 4–6 selected nodes (varying on the bay size) of all five bays. The nodes are selected based on the hierarchy of popularity in the bays, allowing us to select the primary nodes of each bay. The mapping study consists of snapshot headcounts at each node and observation of population density. Based on this mapping study, four levels of very low, low, medium, and high population levels are identified. The second mapping study counts the ICT use for minatory and management uses.

This mapping is conducted simultaneously with the previous mapping study at three times of the day, i.e., morning hours, afternoon hours, and evening hours. This selection is intentional to check and identify different ICT usage apart from CCTV use in all five bays. This mapping is based on three analyses, including headcounts of CCTV usage and identifying other ICT use (such as QR codes, IoT, sensory systems, etc.). The mapping verifies the number of ICT devices and tools and ICT-enabled innovations that may have been used in all five bays throughout the three main operational hours. Four levels of very low, low, medium, and high ICT use (for monitoring and management) are identified based on this mapping study. The third mapping study is to identify and record any illegal fishing activities at three time periods in each season for all five bays. The third step in the study is conducted in two parts. First is on-site interviews of local security and safety officers designated for each bay area. The on-site activities involve interviewing local security and security offices about ICT use, tools and devices, and routine checks. The interviews also verify the level of illegal fishing activities in coastal areas. One interview per allocated time per case is conducted in three site visits. The second part of the analysis is the analysis of official reports, documents, and policies concerning the existence of marine ecological protection and restoration projects. In this part,

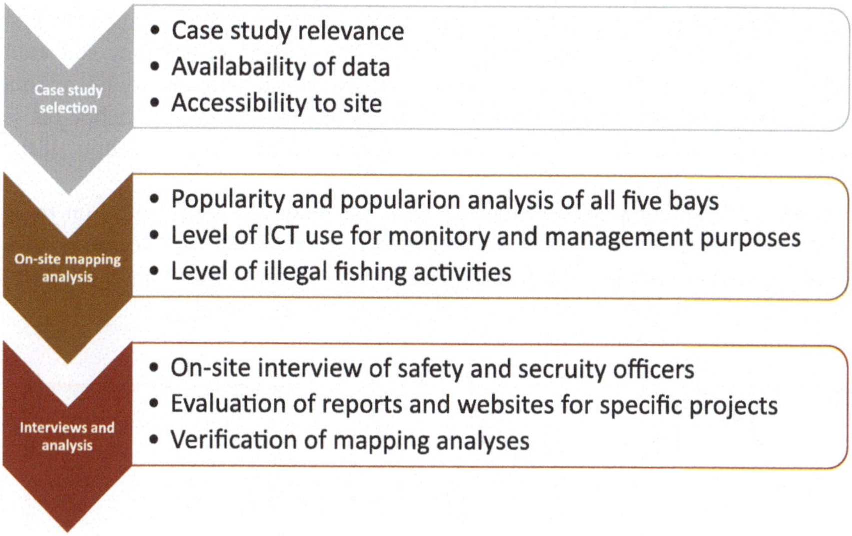

Fig. 4.2 Summary of the study's methods, data collection, and analyses strategy

the analysis is based on local policy document assessment, website assessment, and local community online platforms of each bay area. Figure 4.2 summarises the three steps for data collection and analysis.

4.4 Beach Management: A Case of Five Bays in Sanya, Hainan Island, South China

New bays have been created as a result of Sanya's thriving tourism industry in the last ten or twenty years on Hainan Island, South China. Newly constructed large infrastructure is mostly found around southern coastal areas to support tourism industries as a result of this rapid tourism development. Some of the coastal areas are expanded further away from the city, leading to new tourism activities and destination branding (Chen et al. 2019) and increasing the environmental stress of such regions [38]. As a result of such environmental stress, significant problems have been raised concerning beach pollution, illegal fishery activities, illegal tourism activities, and uncontrolled fishing in coastal areas.

This study is focused on five popular bays, three within Sanya and two adjacent coastal areas to the city. Two out of the three inner coastal areas are primarily popular for mass tourism industries. These are known as 'Sanya Bay', 'Dadonghai Beach', and 'Xiaodonghai Beach'. These are ranked as the busiest coastal areas with a larger influx of people in their beach environments. The third is a smaller public beach, which is less popular with only a few tourism activities and resorts. The other two selected bays are two outer coastal areas of 'Yalong Bay' and 'Sunny Bay'. The first is a public beach area with several larger resorts, hotels, tourism activities, and retail and entertainment units. The second is an exclusive beach area, mainly linked to a few resorts and their facilities. Sunny Bay is not open to the general public but is exclusively used for smaller-scale tourism activities and guests and visitors of the resorts around the bay (Fig. 4.3).

Both local communities and tourists have put more and more strain on these coastal areas in recent years. The region benefits economically from tourism, but it has also brought attention to the difficulties in striking a balance between development

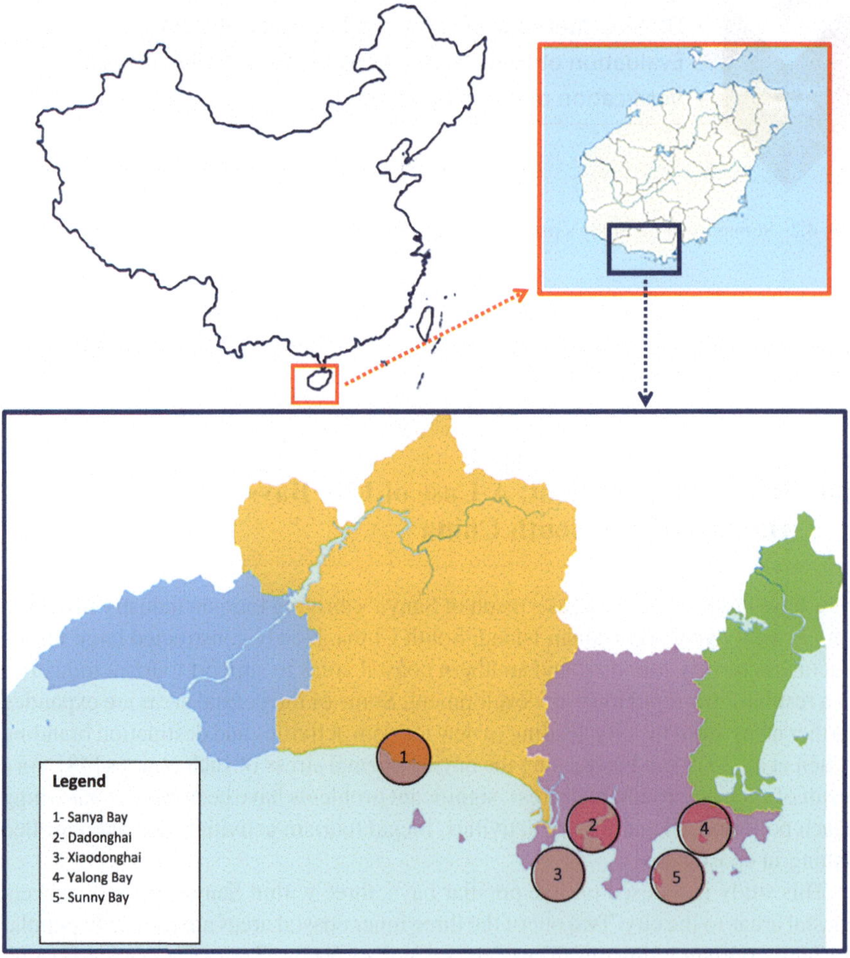

Fig. 4.3 Geographical location of selected five coastal areas as main beaches and bays of Sanya and its adjacent areas, including 1. Sanya Bay, 2. Dadonghai Beach, 3. Xiaodonghai beach (also known as Luhuitou), 4. Yalong Bay, and 5. Sunny Bay. Sanya Bay is located in Tianya District, while the other four area located in Jiyang District of Sanya

and environmental preservation. These problems are especially noticeable in places like Sunny Bay, where restricted tourism and exclusive access help lessen some of the environmental impact, but the quick development of resorts and associated infrastructure nearby causes habitat degradation and biodiversity loss. Both local governments and tourism businesses now prioritize sustainable management due to the continuous strain on the region's infrastructure, resources, and ecosystems. They need to figure out how to lessen the negative effects of excessive tourism while preserving Sanya's appeal as a worldwide travel destination.

Beach management in these five bays is quite different from one another. In two coastal areas of 'Sanya Bay and 'Dadonghai Beach', CCTV cameras are placed alongside the beach areas to detect illegal activities. The facilities and activities are also managed by the local security officers using a sensory system and IoT, which are checked regularly. The sensory system is an outdoor surveillance system, which is supported by CCTV cameras. In other words, the movement and activities are recorded and collected in a centralised system and any unusual activity is reported immediately to notify the on-duty or nearby security and safety officers.

The IoT network is mainly used for maintenance purposes of facilities and report any damages or issues, such as littering, fire, and facility damage. In both areas, coastal fishing activities are limited, while they are overcrowded for other public uses, such as swimming in designated areas, tourism activities, entertainment, etc. There are fewer CCTV cameras in two other coastal areas of 'Xiaodonghai Beach' and 'Yalong Bay'. But in Yalong Bay, a sensory system is also used to support routine checks for any reporting and recording purposes of damages and issues. IoT is currently not present in both bays. Currently, the majority of monitoring in the areas is done by security officers patrolling around regularly around the clock.

In both areas, coastal fishing activities happen more often, which are meant to be illegal. The local security officers verify the lack of ICT use for the purpose of detecting illegal activities, such as illegal fishing. Comparatively, these two areas are less populated, and hence, the limited security measures are also not effective enough. Sunny Bay is an exceptional case out of the five bays, mainly because of its exclusive nature and higher surveillance due to the limited use of beach areas with only restricted designated areas for swimming and limited tourism activities. Hence, coastal fishing activities are absent, and if there are any, they are micro-managed in the bay area. ICT use in this bay is limited to centrally-managed CCTV cameras alongside the coastal areas of the bay. The local officers believe they are sufficient due to less population density in this location. Here, we highlight ICT usage in all five areas (See Figs. 4.4, 4.5, 4.6, 4.7, and 4.8):

1. **Sanya Bay:** CCTV cameras, Sensory system for surveillance, and IoT for maintenance purposes.
2. **Dadonghai Beach:** CCTV cameras, Sensory system for surveillance, and IoT for maintenance purposes.
3. **Xiaodonghai Beach:** CCTV cameras (only).
4. **Yalong Bay:** CCTV cameras, and Sensory system for surveillance.
5. **Sunny Bay:** CCTV cameras (only).

Fig. 4.4 The long stretch of Sanya Bay between Sanya's CBD and Sanya Airport. *Source* Generated by the author using the Google Earth open access maps

Fig. 4.5 The stretch of Dadonghai Beach in south-eastern part of Sanya. *Source* Generated by the author using the Google Earth open access maps

Fig. 4.6 The cozy stretch of Xiaodonghai Beach in southern part of Sanya. *Source* Generated by the author using the Google Earth open access maps

Fig. 4.7 The long stretch of Yalong Bay in the eastern outskirts of Sanya. *Source* Generated by the author using the Google Earth open access maps

Fig. 4.8 The small stretch of Sunny Bay in the eastern outskirts of Sanya. *Source* Generated by the author using the Google Earth open access maps

(See Figs. 4.9, 4.10, 4.11, 4.12, and 4.13 for all five cases).

Table 4.1 summarises some of the main findings from conducted mapping surveys, interviews, and analysis of reports, websites, and official documents, related to the five bays in Sanya.

Although it is limited, the mapping analysis of these five coastal areas above shows a strong correlation between coastal fishing activities and the use of ICT-enabled innovation. We observe less illicit fishing in coastal areas in the two cases of Sanya Bay and Dadonghai Beach, where there is a range of ICT use beyond CCTV cameras. This is supported more by the sensory surveillance system installed at both bays, but it is also partially influenced by the increased level of security and tourism. In addition to Sunny Bay's exclusive setting, ICT use has improved beach and bay area management and monitoring. This primarily relates to illicit fishing and coastal fishing, which could be curbed with the use of sensors, CCTV cameras, and real-time data gathering. We observe a greater degree of illicit fishing in coastal areas in two cases, Xiaodonghai Beach and Yalong Bay, where ICT use is restricted. There are only a few officers present in both situations. In order to identify and document any illicit fishing activities, the local security and safety officers thus mainly rely on their regular inspections. Notably, neither Xiaodonghai Beach nor Dadonghai Beach currently have any projects aimed at protecting and restoring the marine environment.

Furthermore, in terms of both quantity and variety, ICT use is more prevalent in the more well-known areas of Dadonghai Beach and Sanya Bay. The outdoor surveillance system, which is run by an embedded sensory network to identify and

Fig. 4.9 The presence of beach management, security, and CCTV cameras (in distance in the right) in Sanya Bay. *Source* The Author's own

document any unlawful activity, is supported by the CCTV cameras in both locations. Officers on duty and/or in the vicinity can look more closely and respond to any illegal activity more quickly thanks to the real-time data collection, especially in the early morning, evening, and night when fewer people are around. The biggest obstacle, though, still exists in the evening and at night, when there is little to no light to aid in prompt verification and reaction. As a result, there are more local security and safety officers on duty during those hours, and their duties extend beyond spotting illicit fishing. Additionally, they regularly check for other illegal activities like drug use and trafficking, social misbehaviour, damage, littering, illegal sales, barbecuing, etc. Examples of CCTV use and notifications are shown in Figs. 4.14, 4.15, and 4.16

Despite these efforts, the effectiveness of the surveillance system is still limited by challenges such as the need for greater lighting infrastructure and the high demand for human resources during off-peak hours. Future improvements in both technology and human resource allocation are necessary to enhance the system's capacity to ensure greater security and promote responsible tourism in these areas.

Fig. 4.10 The presence of night security patrols and CCTV cameras in Dadonghai Beach. *Source* The Author's own

4.5 Further Discussions on Key Findings

Here, we debate whether ICT-enabled innovation is an impactful tool for change in beach management. Undoubtedly, we could learn from the case of multiple coastal areas (i.e., beaches and bays) in Sanya, the number one beach destination in Mainland China. The correlation between the presence of ICT tools and better monitory of the coastal fishing activities is inevitable. But why do we still hesitate to use such tools for beach management strategies? If financial constraints are not the problem, then there should be no barriers to promoting ICT-enabled innovation for beach management.

The fish population of populated beach areas, like Sanya, is decreasing drastically, which is alarming. Over the years, the fishing intensity has grown mainly due to booming tourism industries and associated activities in coastal areas, such as mass tourism, fishing, and entertainment. The coastal regions are now dealing with extreme pollution, illegal tourism activities (such as unofficial water sports activities, deep water fishing, etc.), and uncontrolled coastal fishing activities. The latter is of concern for Sanya and places alike, as they are often not included in the ecological or environmental protection strategies. Based the [28] conceptual model, ICT could be a practicable tool to overcome the limitations in coastal or beach management strategies. ICT usage could make a difference, and some could occur with simple ICT-enabled innovations and interventions. While limited to security and monitory

Fig. 4.11 The presence of informal tourism and fishery in Xiaodonghai Beach. *Source* The Author's own

measures, the current practices could develop further to raise awareness, have better knowledge about coastal activities, and minimise human impact via (illegal) coastal fishing activities. The suggestions are to enable the development of sensory networks and real-time data collection to ensure coastal fishing activities are abolished. In doing so, beach management systems must include ICT-enabled models to improve monitory measures, promote environmental protection, and develop new policies against illegal fishing activities in the coastal areas.

One of the study's main conclusions emphasizes the need for a more thorough and proactive ICT-enabled governance framework that substitutes dynamic and predictive management for passive monitoring. The wider environmental and social issues brought on by conflicting coastal activities cannot be adequately addressed by the current reliance on static surveillance systems, such as CCTV cameras. According to the study, illegal fishing activities have significantly decreased in areas with integrated ICT use, like Sanya Bay and Dadonghai Beach, while areas with limited ICT deployment, like Xiaodonghai Beach and Yalong Bay, still face difficulties with regulatory enforcement. This discrepancy shows that, when properly combined with enforcement and policy frameworks, ICT tools not only improve surveillance but also act as a deterrent. But the true promise of ICT is found in its ability to produce useful insights through predictive analytics, community engagement platforms, and real-time environmental data collection. Beach management tactics can move from

Fig. 4.12 The presence of guards in designated zones alongside ICT-based interventions in Yalong Bay. *Source* The Author's own

reactive enforcement to preventive governance by utilizing IoT-based monitoring systems, automated enforcement alerts, and GIS mapping for trends in coastal activity. Furthermore, by offering transparent, data-driven decision-making tools that guarantee equitable resource distribution while addressing ecological concerns, ICT can promote stakeholder collaboration. The results of the study demonstrate that ICT is not only a supporting tool but also a basic requirement for contemporary beach management, as it facilitates flexible, fact-based policymaking and, in the end, ensures the long-term viability of Sanya's coastal resources.

4.5.1 Recommendations

An integrated governance approach is necessary to limit harmful near-coast activities and maintain the long-term environmental quality of coastal recreational resources. First, to clearly define the limits of what is allowed and what is not, a mix of ICT-enabled monitoring systems and regulatory policies should be put into place. For instance, automated reporting systems and real-time surveillance can help enforce no-fishing zones and restrict illicit activity, maintaining the ecological sustainability of recreational areas. Furthermore, adaptive zoning techniques can reduce

Fig. 4.13 Sunny Bay may appear as an example of a more private beach or bay areas in compared to other four cases. *Source* The Author's own

Table 4.1 Summary of coastal area/bay/beach popularity in correlation with the level of ICT use and illegal coastal fishing activities in five bay areas in Sanya

	Popularity and population level in coastal areas	Level of ICT use for monitory and management	Level of illegal fishing activities in coastal areas	Marine ecological protection and restoration projects*
Sanya Bay	High	High	Low	Yes
Dadonghai Beach	High	High	Low	No
Xiaodonghai Beach	Low	Low	High	No
Yalong Bay	Medium	Medium	High	Yes
Sunny Bay	Very Low	Low	Non-existent	No

* *NOTE* Most recent information is extracted from Sanya government website (see http://english.sanya.gov.cn/syen/news/202203/9884e019e0dc45df88fc3a1569f21c3d.shtml) updated on 28 March 2022

Fig. 4.14 The use of CCTV cameras in busier beach and bay areas have become a common practice. Here is an example of Houhai Village bay slightly further away from Sanya. *Source* The Author's own

Fig. 4.15 Example of notifications for video monitoring zones, 24 h surveillance, and CCTV usage in Xiaodonghai Beach. *Source* The Author's own

Fig. 4.16 Another Example of notification for visitors and residents indicating 24 h surveillance, and CCTV usage in Xiaodonghai Beach. *Source* The Author's own

conflicts between conservation and recreation objectives. Examples of these include areas designated for responsible tourism and fishing, or seasonal restrictions on specific activities. Authorities can enable data-driven decision-making and create dynamic regulations that quickly address new environmental threats by utilizing ICT platforms.

Second, to encourage sustainable coastal recreation, enforcement mechanisms can be strengthened by community involvement and incentive-based conservation initiatives. By educating tourists, locals, and stakeholders about ecological preservation initiatives, digital platforms can promote voluntary adherence to environmental regulations. Additionally, for communities that depend on near-coast activities for their livelihoods, economic incentives like ecotourism opportunities and sustainable fisheries certification programs can offer viable alternatives. Policymakers can develop a balanced strategy that protects coastal ecosystems while making sure that businesses and local communities actively contribute to the environmental resilience of the area by combining ICT tools with participatory governance models.

ICT must be created as a proactive tool for change in order to revolutionize beach management in Sanya. This will involve combining stakeholder collaboration, predictive analytics, and real-time data collection into a single system. The results of the study show that ICT-enabled solutions, like AI-driven analytics for tracking ecological and human interactions, geofencing for restricted areas, and automated environmental sensors, can significantly improve resource use balance. For instance, automated alerts for law enforcement agencies coupled with real-time tracking of illicit fishing activities via IoT-connected sensors could greatly increase compliance while lowering the need for human patrols. In order to promote voluntary compliance and lessen the need for punitive enforcement measures, mobile applications and digital engagement platforms can also be used to inform visitors and local communities about environmental regulations. Policymakers can move away from dispersed enforcement and toward a coordinated, data-driven governance model that fosters sustainability and resilience by incorporating ICT into a Special Area Management Plan (SAMP), as advised by integrated coastal management (ICM) principles.

Beyond just enforcing the law, ICT plays a crucial role in promoting long-term sustainable development in Sanya's beach and coastal management. Authorities can develop adaptive management techniques that react quickly to changes in the environment and in human activity by utilizing ICT as a governance tool. Seasonal modifications to fishing regulations, restrictions on the number of visitors, and focused conservation initiatives to lessen environmental stress can all be made possible by predictive modelling based on real-time data. Additionally, combining ICT with participatory governance models guarantees that solutions are both efficient and fair. In these models, fishermen, tourism operators, legislators, and the general public collaborate to develop management strategies. The long-term sustainability of Sanya's coastal environment depends on a shift from traditional surveillance to an all-encompassing, ICT-enabled management ecosystem, as evidenced by international case studies. In addition to increasing the effectiveness of beach management, this change will guarantee that Sanya maintains its status as a premier coastal destination while preserving its natural integrity for coming generations.

4.6 Conclusion: Tools for Change?

In order to resolve the conflicts between nearshore water uses and coastal recreation in Sanya in a sustainable manner, a coordinated governance structure that incorporates the opinions of various stakeholders into a single management framework is necessary. The effectiveness of ICT-enabled tools is dependent on the existence of a structured regional management plan, even though they provide an unparalleled ability to gather, examine, and share real-time data on shoreline and nearshore activities. A "special area management plan" (SAMP) would give governments, fishing communities, coastal land users, and the general public the platform they need to jointly define sustainable usage boundaries and enforcement mechanisms, as demonstrated by successful integrated coastal management (ICM) models around the world. ICT data runs the risk of becoming dispersed among agencies with varying priorities in the absence of such a framework, which could result in uneven enforcement and misaligned policies. The authors agree that ICT solutions can shed light on the scope and character of these resource conflicts, but their real contribution is to inform a comprehensive governance approach in which conflicting interests collaborate to create policies that strike a balance between social and economic viability and ecological sustainability. ICT must serve as an enabler within a larger regional planning framework, guaranteeing that enforcement actions are not merely reactive but rather a component of a proactive, integrative, and sustainable management approach. This is the clear lesson to be learned from global coastal governance best practices.

There are serious problems with Sanya's current beach management techniques that jeopardize the long-term viability of its coastal resources. A basic level of monitoring is provided by current methods like CCTV surveillance and on-ground security patrols, but they fall short in addressing the complexity of illicit fishing, environmental degradation, and conflicts between nearshore water uses and recreational activities. The results of the study demonstrate how inconsistent and inadequately integrated ICT use is into a larger management strategy in Sanya's main bays and beaches. ICT-enabled surveillance systems, such as sensory networks and IoT-based monitoring, have helped to reduce illicit fishing in popular tourist destinations like Sanya Bay and Dadonghai Beach. However, the lack of sophisticated ICT frameworks is associated with higher levels of unregulated activities in less developed areas like Yalong Bay and Xiaodonghai Beach. Enforcement becomes reactive rather than preventative due to inefficiencies caused by the dependence on disjointed, site-specific management strategies rather than an all-encompassing digital governance system. The difficulties are further exacerbated by the fact that current Beach Management Units (BMUs), which are designed to control fishing activities, do not integrate real-time data and primarily serve as regulatory organizations without dynamic environmental oversight.

ICTs can only be transformative if we use them wisely. The use of environmental data gathered from multiple sources [17] could be encouraged by an ICT-enabled beach management innovation. This opens up possibilities beyond the use of CCTV cameras, such as sensory systems, improved surveillance systems, Internet of Things systems, real-time networks or data collection, etc. Better use of the data gathered could result from such an approach, which could also raise community awareness and participation [17]. As a result, these platforms might develop into a system that allows citizens, governments, developers, and managers to work together to improve environmental protection for coastal and beach areas. In doing so, a citizen science approach is also promoted [27], which is beneficial to longer-term environmental protection plans. The use of ICT for environmental sustainability [13, 36, 8] is an inevitable pathway that could ultimately increase our attention to critical issues such as the decline of the fish population in populated coastal areas. In reality, extensive human activity typically causes fish populations to decline in beach and coastal areas. However, we may be able to lessen some of these negative effects and the illegal fishing that frequently occurs without proper monitoring and management with ICT-enabled innovation. More significantly, we could create smart communities for sustainable development with the strategic use of ICT [9], giving coastal communities the resources and opportunities they need to have a more sustainable future.

Addressing the issues that coastal communities face, particularly those pertaining to overfishing, illegal fishing, and environmental degradation, could be greatly aided by the integration of ICT-enabled innovations. We can collect data on fish populations, fishing activities, and ecosystem health in real time by using satellite technologies, drones, and smart monitoring systems. Authorities and local stakeholders would be better equipped to enforce rules, identify infractions, and guarantee sustainable fishing methods as a result. Furthermore, ICT solutions can help local populations feel more responsible by promoting community involvement and increasing awareness of conservation initiatives. In addition to safeguarding marine resources, integrating ICT into coastal management gives coastal communities the tools they need to diversify their sources of income, adapt to climate change, and create a more resilient and sustainable future.

In conclusion, it is impossible to overestimate the contribution that ICT makes to the advancement of sustainable development in coastal regions. We can develop a more open, responsible, and robust method of managing coastal resources by utilizing technology. Coastal communities can move from vulnerability to sustainability with creativity, collaboration, and careful planning, which will benefit the environment and future generations. Continuous cooperation between governments, local communities, and the private sector, driven by a common vision of sustainability and social equity, will be necessary to ensure the long-term success of such initiatives.

Acknowledgements The author acknowledges the Ministry of Education, Culture, Sports, Science and Technology (MEXT), Japan Government, and the Network for Education and Research on Peace and Sustainability (NERPS), Hiroshima, Japan.

The author also acknowledges that this chapter is accepted for publication under the tentative title of "'*Tools for Change: ICT-enabled Innovation for Beach Management?*' by the *Journal of Urban Regeneration & Renewal*. The text of this chapter is based on the original submission but is slightly modified to fit the purpose of this book. Permission is sought from the publishers to produce the materials for this chapter. The publishers, *Henry Stewart Publications*, are acknowledged, and I'd like to thank the Journal of Urban Regeneration & Renewal for publishing the work in their Volume 19 in 2025.

References

1. An TG (2021) A plan to operate a beach through safety management prevention using ICT technology. Journal of Convergence for Information Technology 11(12):22–29
2. Ariza E, Sardá R, Jiménez JA, Mora J, Ávila C (2007) Beyond performance assessment measurements for beach management: application to Spanish Mediterranean beaches. Coast Manag 36(1):47–66
3. Botero C, Pereira C, Tosic M, Manjarrez G (2015) Design of an index for monitoring the environmental quality of tourist beaches from a holistic approach. Ocean Coast Manag 108:65–73
4. Cheshmehzangi A (2022) The impacts of ICT platforms on public place use in China. Built Environment 48(2):264–279
5. Cheshmehzangi A, Li HMA (2020) Innovation through urban diversity and achieving comprehensive sustainable urbanism from a community-oriented approach. Curr Urban Stud 8(02):222
6. Cheshmehzangi A, You N, Siri J, Birch E (2024) Harnessing urban innovation to unlock the sustainable development goals. Springer
7. Cheshmehzangi A, Zou T (2024) Literary analysis of the connection between urban innovations and Sustainable Development Goals (SDGs). Sustainability 16(15):6405
8. Cheshmehzangi, A. (2022b). ICT, cities, and reaching positive peace. Singapore: Sringer Nature
9. Cheshmehzangi, A. (2022c). The role of ICT and smartness for peace-building processes and practices. In: ICT, cities, and reaching positive peace. Singapore: Springer Nature Singapore, pp 17–46

10. Cheshmehzangi A, Pourroostaei Ardakani S (2024) Embracing the digital twin paradigm for urban sustainability. In: Pourroostaei Ardakani S, Cheshmehzangi A (eds). Digital twin computing for urban intelligence. Singapore: Springer Nature Singapore, pp 1–11
11. Chesson J, Clayton H, Whitworth B (1999) Evaluation of fisheries-management systems with respect to sustainable development. ICES J Mar Sci 56(6):980–984
12. Cummins V, McKenna J (2010) The potential role of sustainability science in coastal zone management. Ocean Coast Manag 53(12):796–804
13. Dhakal SP (2014) ICT-mediated adaptive capacity of environmental third sector in Australia. J Community Inform
14. Flor AG (2001) ICT and poverty: the indisputable link. In: Third Asia Development Forum on 'Regional Economic Cooperation in Asia and the Pacific', pp 11–14
15. Frampton AP (2010) A review of amenity beach management. J Coastal Res 26(6):1112–1122
16. García M, Veneciano M (2011) Proyectos de ley de costas y desarrollo litoral sostenible desde la óptica geográfica. Contribuciones Científicas GÆA 23:95–107
17. Gouveia C, Fonseca A, Câmara A, Ferreira F (2004) Promoting the use of environmental data collected by concerned citizens through information and communication technologies. J Environ Manage 71(2):135–154
18. Graci S (2013) Collaboration and partnership development for sustainable tourism. Tour Geogr 15(1):25–42
19. Haambiya L, Mussa H, Mulumpwa M (2021) A review on the use of information communication technology (ICT) in fisheries management: A case of Mbenji Island small-scale fishery in Malawi. Afr J Food Agric Nutr Dev 20(7):17113–17124
20. James RJ (2000) From beaches to beach environments: Linking the ecology, human-use, and management of beaches in Australia. Ocean Coast Manag 43(6):495–514
21. Kosamu IB, de Groot WT, Kambewa PS (2017) Actor-based design of a management system for the elephant marsh fishery in Malawi. Soc Nat Resour 30(3):299–314
22. Lukoseviciute G, Panagopoulos T (2021) Management priorities from tourists' perspectives and beach quality assessment as tools to support sustainable coastal tourism. Ocean Coast Manag 208:105646
23. Mahalakshmi P, Vimala DD, Krishnan M, Ravisankar T (2009) Needs assessment of ICT users for implementation of aquacultural development projects in coastal areas
24. Njiru M, Van der Knaap M, Taabu-Munyaho A, Nyamweya CS, Kayanda RJ, Marshall BE (2014) Management of Lake Victoria fishery: Are we looking for easy solutions? Aquat Ecosyst Health Manage 17(1):70–79
25. Ogun LTJ, Storey D, Minnery J (2011) Beyond the beach: Balancing environmental and socio-cultural sustainability in Boracay, the Philippines. Tour Geogr 13(4):549–569
26. Oluoch SJ, Obura DAVID (2008) Assessment of fisherfolk organizations and beach management units (BMU) in the management of fishery resources in Diani-Chale, Southern Kenya. Ten years after bleaching—Facing the consequences of climate change in the Indian Ocean—CORDIO Status Report 1:335–343.
27. Owen RP, Parker, AJ (2018) Citizen science in environmental protection agencies. UCL Press, London, 284–300. Available at: https://discovery.ucl.ac.uk/id/eprint/10066319/
28. Palazón A, López I, Gilart V, Aragonés L, Marcos-Jorquera D, Foti D (2019) New ICT-based index for beach quality management. Sci Total Environ 684:221–228
29. Pourroostaei Ardakani S, Cheshmehzangi A (eds) Digital twin computing for urban Intelligence. Singapore: Springer Nature Singapore
30. Ruddle K (1993) External forces and change in traditional community-based fishery management systems in the Asia-Pacific region. Maritime Anthropological Studies 6(1–2):1–37
31. Sardá R, Valls JF, Pintó J, Ariza E, Lozoya JP, Fraguell RM, Martí C, Rucabado J, Ramis J, Jiménez JA (2015) Towards a new integrated beach management system: The ecosystem-based management system for beaches. Ocean Coast Manag 118:167–177
32. Saville R, Hatanaka K, Sano M, Wada M (2015) Application of information and communication technology and data sharing management scheme for the coastal fishery using real-time fishery information. Ocean Coast Manag 106:77–86

33. Sha'aban NS, Mohd Zahari F (2020) The usage of ICT to raise knowledge and awareness about beach cleanliness among community at Pantai Merdeka. In: Symposium on Technology Management and Logistics (STML Go Green) 2019: Management of Technology, Volume 1 (Vol. 1, No. 25, pp 61–62). Universiti Utara Malaysia.
34. Shipman B, Stojanovic T (2007) Facts, fictions, and failures of integrated coastal zone management in Europe. Coast Manag 35(2–3):375–398
35. Thirumaran K, Raghav M (2017) Marketing sustainable beach holidays: a tropical agenda. In: Balancing development and sustainability in tourism destinations. Springer, Singapore, pp 401–409
36. Wijeyaratne MJS (2012) Use of ICT for sustainability: case of a new middle-income country after a long period of armed conflict. International Conference on ICT and Sustainability Science, Ibaraki, Japan
37. Yamamoto T (1995) Development of a community-based fishery management system in Japan. Mar Resour Econ 10(1):21–34
38. Zou X, Wang Y, Wang X, Ge C, Niu Z (2004) Environmental stress and policy guidelines for environmental management of the Sanya coastal zone, Hainan Island, China. J Coast Res: 243–253.

Chapter 5
Innovation and Social Sustainability: ICT Integration in Residential Compounds for Empowering Elderly Communities

5.1 Introduction: Context and Rationale

The southernmost point of Hainan Island, Sanya, is going through a major demographic transition, with a growing number of senior citizens. The 2020 census shows that there are 1,031,396 people living in Sanya, 10.42% of whom are 60 years of age or older, and 6.95% of whom are 65 years of age or older [27]. Interestingly, the percentage of residents 60 and older has increased by 3.41% since the 2010 census, whereas the percentage of residents 65 and older has increased by 2.25%. The general aging trend seen in China is reflected in this demographic trend. By 2019, there were 254 million people in the country who were 60 years of age or older, by 2040, it is predicted that this number will increase to 402 million, or 28% of the total population. The influx of elderly migrants, sometimes known as "migratory birds," who temporarily move to Sanya due to its ideal climate and health advantages [18], contributes to the city's aging population. The region's growing elderly population is a result of this seasonal migration [22].

There are a number of issues with Sanya's aging population, especially in the suburbs. The burden on healthcare services is one important problem [21]. Even though Hainan Province has made progress in building senior care facilities—as of late 2022, there were 135 operational institutions—rural areas continue to face obstacles. 96.2% of these facilities provide medical services, and they include 59 senior-only canteens and 203 community daycare centers. Despite these efforts, the rapid increase in the elderly population necessitates further expansion and enhancement of healthcare infrastructure to meet the growing demand.

The prevalence of noncommunicable diseases (NCDs) in the elderly is another urgent issue [2, 17]. According to national data, NCDs like diabetes, hypertension, and cardiovascular diseases affect about 75% of China's senior citizens [15]. Given that Sanya's suburban areas may have less concentrated healthcare resources, the high prevalence of these chronic conditions highlights the need for easily accessible and specialized medical care. Another issue facing the elderly population is social

A. Cheshmehzangi, *Urban Innovation and Sustainability*, Urban Sustainability,
https://doi.org/10.1007/978-981-96-6370-5_5

isolation, particularly in suburban and rural areas [31]. Older adults in rural areas are isolated and segregated as a result of factors like smaller families and greater job mobility.

This seclusion can negatively impact mental health and general well-being, highlighting the significance of support systems and community involvement [32]. Sanya's suburban areas' elderly population's well-being is also greatly influenced by economic factors. Many elderly people in rural areas struggle with insufficient pensions and medical security, which makes them even more vulnerable. A lower quality of life can result from this financial instability, which can also restrict access to essential healthcare services [19]. A potential remedy for these issues is the incorporation of ICT into residential complexes. Elderly' access to healthcare can be improved by ICT-enabled telemedicine services [13], which allow for remote consultations and ongoing health monitoring [5]. Additionally, by linking seniors with friends, family, and local resources, digital platforms can encourage social interaction and lessen feelings of loneliness. Smart home technology can also improve convenience and safety, enabling senior citizens to continue living independently while guaranteeing timely help when needed.

In sum, the suburban areas of Sanya are experiencing a demographic shift towards an aging population, which poses a number of complex issues such as social isolation, economic insecurity, chronic disease management, and healthcare accessibility [7]. A workable solution to these problems is the strategic integration of ICT in residential settings, which promotes social sustainability and innovation while empowering the area's elderly populations.

5.1.1 ICT and Social Sustainability: The Role of ICT in Promoting Social Sustainability and Well-Being for Elderly Communities

Information and communication technology (ICT) integration into residential complexes is essential for advancing social sustainability and improving senior citizen communities' quality of life. This strategy tackles the difficulties brought on by a rapidly aging population and the resulting social and health ramifications in the Sanya suburbs of Hainan Island, China. We can gain important insights into practical methods for empowering senior citizens in Sanya by looking at successful ICT implementations both domestically and internationally.

Social sustainability involves creating communities that are equitable, diverse, connected, and democratic, providing a good quality of life [1, 16]. For elderly populations, ICT serves as a catalyst in achieving these objectives by (Fig. 5.1):

1. **Facilitating Social Inclusion**: ICT platforms enable older adults to maintain connections with family, friends, and the broader community, thereby reducing feelings of isolation.

Fig. 5.1 Summary of ICT as catalyst across multiple objectives of social sustainability

2. **Providing Access to Services**: Digital tools offer convenient access to healthcare, financial, and recreational services, enhancing the autonomy and quality of life for the elderly.
3. **Promoting Lifelong Learning**: Online educational resources empower elderly individuals to engage in continuous learning, fostering cognitive health and personal fulfillment.

ICT integration in suburban Sanya residential complexes can help the area address important socioeconomic and infrastructure issues that impair senior citizens' quality of life. Due to demographic shifts and geographic dispersion, many seniors in these areas experience social isolation, limited mobility, and inadequate healthcare facilities. By offering real-time support and connectivity, smart technology solutions like digital communication platforms, telemedicine services, and AI-driven health monitoring can close these gaps [4]. Additionally, by encouraging a sense of engagement and belonging, community-based digital initiatives can assist elderly in becoming more involved in peer-to-peer support networks, cultural events, and local governance [8]. By maximizing resource allocation, easing the strain on physical infrastructure, and facilitating more effective service delivery, ICT integration enhances the resilience of aging communities in addition to providing personal advantages. Given the rapid urbanization and ongoing efforts to modernize Sanya's suburban areas [11], leveraging ICT as a tool for social sustainability will not only enhance elderly residents' quality of life but also set a precedent for other regions facing similar demographic transitions.

The suburban areas of Sanya have unique problems, such as inadequate social services for the elderly and healthcare infrastructure. These problems are made worse by the seasonal influx of "migratory bird" retirees, who force the local systems to adapt to a changing elderly population. These difficulties can be lessened by implementing ICT solutions in these areas by: (1) Enhancing Healthcare Delivery: Telemedicine platforms can bridge the gap between healthcare providers and elderly patients, offering remote consultations and continuous health monitoring; (2) Strengthening Community Networks: Social media and community forums tailored for seniors can foster a sense of belonging and mutual support among residents; and (3) Improving Accessibility: User-friendly interfaces and localized content ensure that ICT services are accessible to all elderly individuals, regardless of their technical proficiency.

Many effective ICT projects around the world provide insightful guidance for incorporating technology into elderly care in Sanya's suburbs. The Beihang Elderly Care Alliance in China has created an intelligent community-based care service model that incorporates ICT to manage medical records, track health metrics, and help senior citizens and healthcare providers communicate. Seniors in urban areas now have much better access to and quality of healthcare thanks to this system. Another illustration is Yunlin Life Apartments, a complete smart home care system that uses sensors and Internet of Things gadgets to keep an eye on senior citizens' everyday routines and health. Seniors living alone now feel safer and more independent thanks to this program, which provides automated emergency responses and real-time health alerts.

Other international projects outside of China demonstrate the potential of ICT in elderly care. An AI-powered service in South Korea called CLOVA AI CareCall regularly calls elderly people who live alone to check on their wellbeing and offer any assistance that may be required [20, 25]. This program has successfully decreased social isolation and made sure that seniors receive timely support. In the meantime, Japan's Digital Garden City Nation Vision program encourages the use of digital technologies to support aging populations and revitalize rural areas. This initiative has made communities more sustainable and elderly-friendly [10] by implementing ICT infrastructure to improve daily living, healthcare, and transportation.

These case studies suggest a number of approaches for incorporating ICT into elderly living communities in the suburbs of Sanya. The creation of intelligent healthcare systems, such as remote health monitoring and telemedicine, would offer ongoing care, especially for people with long-term illnesses. While guaranteeing seniors receive regular medical care, these digital health solutions can lessen the strain on physical healthcare facilities.

Seniors should also be introduced to ICT tools through digital literacy programs so they can make the most of the resources that are available to them [9]. Providing structured training sessions can enable older adults who may find it difficult to use digital platforms to confidently navigate new technologies. Online forums and social networks catered to senior citizens' interests and needs should be created in order to further strengthen community ties by encouraging peer support and social engagement. Another important tactic is cooperation with technology suppliers. Sanya can encourage the creation of user-friendly gadgets and apps that are tailored for senior citizens by collaborating with tech firms. ICT tools can be made more accessible with features like voice commands, streamlined navigation, and large fonts. In order to safeguard the private data of senior citizens and promote confidence in digital solutions, it is imperative that robust data privacy and security measures be put in place.

5.1.2 ICT Integration in Suburban Sanya

The suburban areas of Sanya have unique problems, such as inadequate social services for the elderly and healthcare infrastructure. These problems are made worse by the seasonal influx of "migratory bird" retirees, or elderly people from other regions of China who spend the winter in Sanya. This is because the local systems have to adapt to a changing elderly population. By enhancing healthcare access, building community ties, and guaranteeing digital inclusivity, the use of ICT solutions can help lessen these difficulties.

By providing remote consultations and ongoing health monitoring, telemedicine platforms can help close the gap between senior citizens and healthcare professionals. Telemedicine guarantees prompt access to medical advice and treatment because many elderly people have mobility issues or reside in places with few healthcare facilities. Additionally, enhancing community networks via ICT-based social media platforms can help senior citizens feel like they belong. Senior-focused digital community groups and online forums can help fight loneliness and offer a place for support and shared experiences. Furthermore, enhancing accessibility through localized content and user-friendly interfaces guarantees that all senior citizens, irrespective of their level of technical expertise, can make effective use of these digital tools.

ICT integration has many advantages, but successful implementation requires addressing a number of issues. The digital divide is a significant barrier because older people's technological proficiency varies [12]. Many people may find it difficult to use digital services if they are not properly trained. To close this gap, specialized educational initiatives like workshops and one-on-one support will be crucial. Limitations in suburban areas' infrastructure present another difficulty. To enable widespread ICT adoption, it is necessary to guarantee dependable internet connectivity, enough hardware, and continuous technical support. Many seniors won't be able to access even the most cutting-edge digital services without robust infrastructure. Another factor to take into account is cultural resistance to new technology. Because they are unfamiliar with digital solutions or are worried about security, some seniors may be reluctant to embrace them. Initiatives for community engagement that highlight the concrete advantages of ICT for everyday convenience, healthcare access, and social connectivity should be put into place to promote acceptance.

In suburban Sanya, integrating ICT into residential compounds offers a practical way to improve elderly communities' well-being and social sustainability. Sanya can create customized strategies that tackle its distinct demographic and regional issues by taking inspiration from successful domestic and international initiatives.

Accessibility-focused solutions, community-building platforms, and digital healthcare have the potential to revolutionize elderly care in the area. Such initiatives will ensure long-term social sustainability for Sanya's aging population by enhancing the quality of life for elderly communities and fostering a more cohesive and resilient community.

5.1.3 Setting the Scene: Research Aims and Objectives

In order to improve quality of life, increase social sustainability, and promote digital inclusion among the elderly, the main goal of this research work is to examine how Information and Communication Technology (ICT) is integrated into elderly communities in four suburban and town areas of Sanya's Tianya District: Liuxiang Town, Jiucai Village, Shuijiao Village, and Huihui Community (Fig. 5.2). The goal of this researchwork is to determine how ICT advancements can alleviate socioeconomic difficulties, rectify infrastructure shortcomings, and design age-appropriate spaces that are suited to the unique requirements of senior citizens living in these communities.

Senior citizens in Tianya District, especially in its western and northwest regions, frequently experience social isolation, healthcare constraints, and economic instability due to the district's mix of established town areas and urbanizing rural settlements. Due to high rates of outmigration among younger generations looking for work in central Sanya or elsewhere, Liuxiang Town, the largest of the four study areas, is experiencing a rapidly aging population. Due to limited access to professional elderly care services, many seniors continue to rely on unofficial networks of caregivers. Due to dwindling agricultural productivity and a lack of alternate sources of income, older residents of Jiucai Village, a primarily agricultural settlement, face severe financial hardships. The elderly population of Shuijiao Village, a coastal community, is impacted by seasonal employment patterns, a lackluster healthcare system, and restricted access to new technologies. The elderly population of Huihui Community, which has recently experienced urbanization efforts, is diverse and includes both local seniors and new retirees from other regions of China, especially during the winter. Yet, despite being relatively close to cities, digital literacy rates are still low, which limits older people's ability to take full advantage of ICT-driven services.

The study intends to evaluate how ICT-driven solutions can address basic issues like elderly social isolation, economic vulnerability, and limited access to healthcare within these four communities. ICT integration has the potential to revolutionize how senior citizens interact with their surroundings and obtain necessary services through the use of telemedicine services, digital community networks, and smart home technologies. For example, telehealth platforms that link elderly patients in remote areas with medical professionals in central Sanya can greatly improve healthcare accessibility. Because chronic conditions like diabetes and hypertension are so common

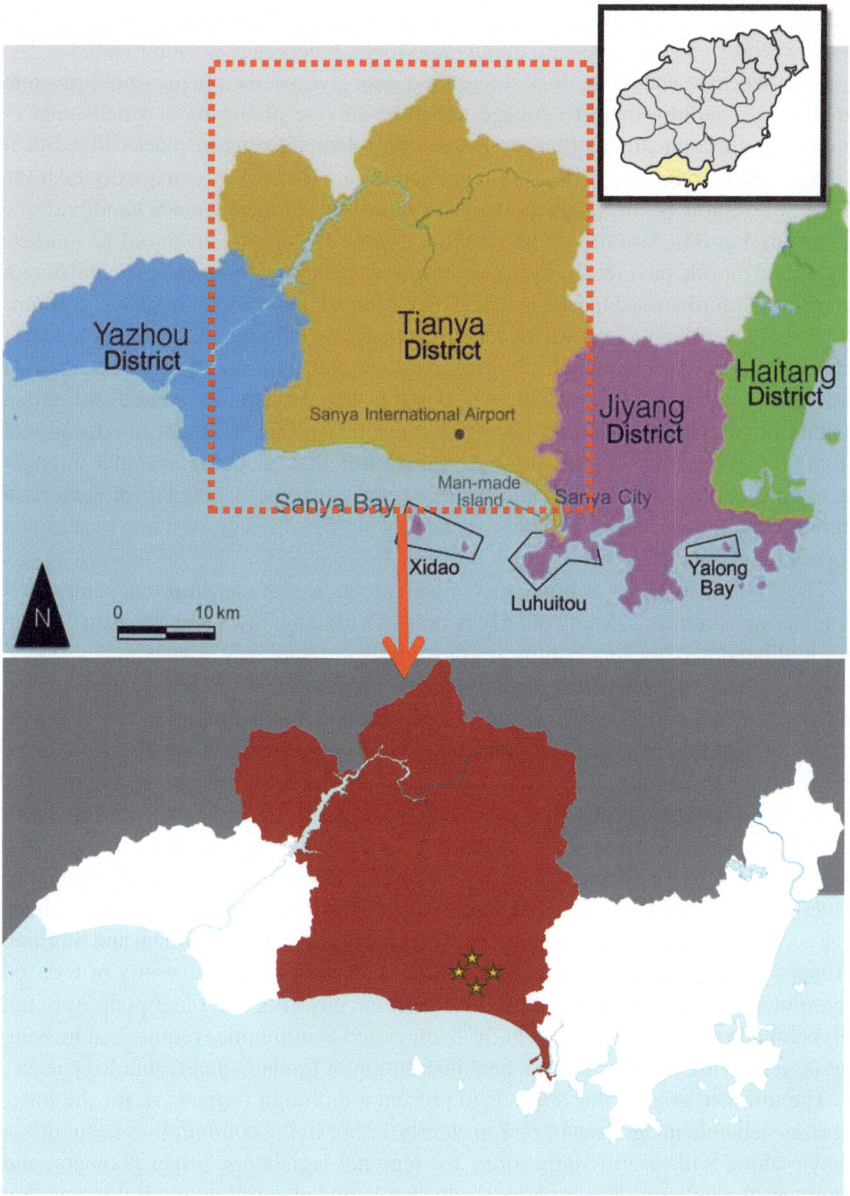

Fig. 5.2 Location of four case samples within the Tianya District of Sanya, Southern Hainan, China. *Source* maps are adapted from various sources

in senior citizens, real-time health monitoring via wearable technology and digital health records can improve care quality and lower emergency hospital visits.

Economically speaking, ICT integration can give senior citizens new revenue streams by enabling them to engage in online service platforms or small-scale e-commerce through digital literacy initiatives. Older citizens in places like Jiucai Village, where traditional agricultural livelihoods are dwindling, can use digital tools to conduct remote business operations like online sales of locally made handicrafts or agricultural goods. Training initiatives that expose senior citizens to online marketplaces and mobile payment systems can promote economic independence and lessen reliance on family assistance. Since ICT integration can promote increased community engagement and lessen the psychological effects of aging alone, social sustainability is a crucial component of this research. Digital community networks, created especially for senior citizens in these towns and suburbs, can offer venues for online social interactions, access to neighborhood events, and involvement in educational programs. For example, structured digital forums that encourage regular engagement with both local residents and visiting retirees can be beneficial for seniors in Huihui Community, who frequently struggle with social integration due to seasonal population shifts.

The study also looks at how smart home technologies can improve senior citizens' living conditions. Many elderly people in Liuxiang Town and Shuijiao Village live in dilapidated housing complexes devoid of contemporary conveniences, which makes day-to-day chores more challenging as people age. Seniors can continue living independently for longer periods of time by implementing smart security features, automated lighting systems, and voice-activated home assistants, which can increase convenience and safety. ICT-based safety monitoring systems can also help caregivers by giving them up-to-date information on senior citizens' health and enabling prompt action in the event of an accident or medical emergency.

Investigating how Tianya District's digital infrastructure might be developed to better serve the needs of senior communities is one of the study's other main goals. The accessibility of ICT-based services is limited in some areas of Jiucai and Shuijiao Villages due to limited internet connectivity, underscoring the necessity of focused investment in broadband expansion. Furthermore, ensuring that older people are not left behind in the shift to intelligent, ICT-integrated communities requires addressing the digital divide by offering practical instruction in fundamental technology use.

The ultimate goal of this study is to present a thorough framework for the long-term, sustainable integration of ICT in elderly/senior living communities in suburban Sanya, along with useful suggestions for regional legislators, urban planners, and community members. The study will advance knowledge of how digital innovation can promote social sustainability, empower senior citizens, and improve the general standard of living in these changing suburban settings by concentrating on the distinct demographic and socioeconomic circumstances of Liuxiang Town, Jiucai Village, Shuijiao Village, and Huihui Community.

5.2 Residential Compounds and Elderly Living in Sanya

For elderly communities of Sanya's suburban residential areas in Tianya District, namely Liuxiang Town, Jiucai Village, Shuijiao Village, and Huihui Community, were tested as places where ICT integration was experimented as part of a public engagement exercise in 2022. These cases present unique but connected challenges. Even though these regions profit from Hainan's overall urbanization and tourism-driven economic expansion, they still lack a significant amount of digital infrastructure and ICT-based public services. Important concerns about healthcare access, financial security, and social well-being for the elderly population may be resolved by integrating ICT into these suburban communities. This study investigates how ICT-driven innovations in residential living environments, public spaces, and public services can raise the standard of living for the elderly and promote social sustainability by looking at these four compounds.

Both traditional and changing features can be found in Sanya's suburban residential areas, especially in Tianya District. Elderly people usually reside in multigenerational households or aging independent homes in Liuxiang Town, a regional center with a mix of new and old housing structures. Jiucai and Shuijiao Villages, where older people frequently reside in rural areas with little access to digital services, stand in contrast to the town's comparatively advanced public infrastructure. As younger generations move to cities, these villages suffer from growing depopulation despite maintaining their traditional community structures. Conversely, Huihui Community is a transitional suburban area where recent mainland Chinese retirees move in for the season, resulting in a mix of long-term residents and short-term senior citizens. Despite having comparatively better facilities, older populations in Huihui still have low levels of digital literacy, which restricts their use of ICT-based services.

Throughout these four compounds, socioeconomic and cultural factors are important in determining the lifestyles of the elderly. Because local residents' retirement benefits are frequently insufficient, particularly in agricultural communities like Jiucai and Shuijiao, many seniors in these areas suffer from financial dependency. Huihui Community, on the other hand, has an older population that is more financially secure, but because of their erratic lifestyles, these retirees frequently have difficulty integrating into society. In Liuxiang Town, where family members have historically taken care of elderly relatives, the strong Confucian value of filial piety continues to influence elder care. However, this custom has been broken by younger generations moving to cities in search of work, leaving many elderly people without sufficient family support. Furthermore, the "migratory bird" retiree phenomenon—in which senior citizens from northern China spend a portion of the year in Sanya—has created a new dynamic in places like Huihui Community, where seasonal population fluctuations pose a threat to the long-term viability of community networks. By developing digital community platforms that allow local and seasonal senior citizens to interact, exchange resources, and access support systems, ICT can aid in closing these social divides.

These suburban areas' current community infrastructure offers both possibilities and limitations for ICT-based interventions. Compared to Jiucai or Shuijiao Villages, where senior citizens must travel great distances for specialized care, Liuxiang Town offers comparatively easier access to healthcare services. Seniors can consult doctors remotely without having to travel, thanks to telemedicine services, which can be a game-changer. In order to give caregivers and medical professionals real-time health updates, digital health monitoring systems, such as wearable technology that tracks vital signs, can be deployed in all of these areas. Additionally, if placed in residential complexes, AI-driven emergency response systems can guarantee that senior citizens receive prompt assistance in the event of an accident or medical emergency.

ICT can improve the functionality of public spaces in these suburban areas, which are still largely underutilized for senior-friendly activities. Smart kiosks and interactive digital boards can inform senior citizens about health programs, community events, and public services in Liuxiang Town, where parks and community centers are the main places for people to socialize. Seniors can express their concerns and participate in local decision-making by using these digital hubs to enable virtual participation in governance initiatives. Virtual reality (VR)-assisted exercise programs can be implemented in Shuijiao Village, where coastal areas could be transformed into recreational areas, to promote physical activity among elderly people who might have mobility challenges. Seniors can participate in customized exercise regimens based on their medical conditions by integrating AI-powered fitness guidance systems in these public spaces, which will enhance their general well-being.

Smart home technologies can also improve the security and comfort of senior citizens in residential areas of these communities. The lack of contemporary conveniences in many homes in Jiucai and Shuijiao Villages makes life challenging for elderly residents. Senior independence and safety can be greatly increased by putting in place sensor-based security features, automated lighting systems, and voice-activated home assistants. For instance, stoves can be programmed to automatically turn off when left unattended, preventing fire hazards. Furthermore, remote home monitoring systems can help retirees in Huihui Community, where a large number of them own seasonal homes, control environmental conditions and household security from their mobile devices, guaranteeing a smooth transition between their stays in Sanya and other regions of China.

ICT promotes digital inclusion and long-term social sustainability in addition to providing immediate functional improvements in public services, public areas, and residential living. Programs for digital literacy geared toward senior citizens must be implemented in community centers, medical facilities, and local marketplaces to guarantee a smooth integration. In order to lower adoption barriers among seniors with little technological experience, these training programs should concentrate on making ICT tools accessible and easy to use. Elderly people can participate in local digital marketplaces and remote financial transactions by working with local companies and service providers to promote the adoption of digital payment methods and e-commerce platforms.

In conclusion, there is a revolutionary chance to rethink elderly living in suburban Sanya through ICT interventions in Liuxiang Town, Jiucai Village, Shuijiao Village, and Huihui Community. These fields can overcome social isolation, financial constraints, and healthcare disparities by utilizing telemedicine, smart home innovations, AI-powered social engagement tools, and digital community networks. In addition to improving the quality of life for senior citizens, ICT-based public services, enhanced public areas, and intelligent residential settings will set a long-term precedent for Sanya's aging-friendly suburban communities. Building a socially sustainable future will depend on the city's aging population being integrated, empowered, and supported by digital technology as it develops.

5.3 From Experimentation to Reality: ICT Integration for Social Sustainability

The integration of ICT into elderly communities in our four samples locations of Liuxiang Town, Jiucai Village, Shuijiao Village, and Huihui Community represents both an experimentation process and a transformative step toward achieving social sustainability. These Sanya suburbs are ideal locations for cutting-edge ICT applications because of the substantial demographic changes, problems with healthcare accessibility, and social isolation they face. These communities can test, improve, and put into practice solutions that will improve the quality of life for the elderly population by piloting digital platforms, smart home technologies, and community-centered ICT initiatives. In order to move from experimentation to reality and establish a model for wider adoption in Sanya and beyond, this initiative draws inspiration from successful examples in other urban and suburban contexts.

The absence of easily accessible healthcare services for senior citizens is one of the most urgent issues in these communities. Residents of Jiucai and Shuijiao Villages must travel great distances for medical consultations due to the underdeveloped healthcare system in Liuxiang Town. Telemedicine platforms and smart home health monitoring systems, which have been successfully tested in cities like Guangzhou and Hangzhou, can help close this gap. Seniors can perform remote examinations and virtual consultations with physicians during this experimental phase by installing digital health kiosks in community centers and nearby clinics. To help medical professionals identify early indicators of chronic diseases and ensure prompt medical interventions, these kiosks can be integrated with AI-driven diagnostic tools.

Smart home technologies can greatly enhance senior care in residential settings [3, 14, 24]. In Jiucai and Shuijiao Villages, a large number of elderly people live alone or in traditional homes devoid of contemporary safety features. Installing voice-activated gadgets, automated home systems, and AI-powered sensors can make senior citizens' living spaces safer and more comfortable. For instance, automated medication dispensers can remind residents to take their prescribed medication on time, and motion-detection sensors in homes can notify caregivers if an elderly person falls. By offering real-time weather, appointment, and emergency contact information, voice-activated home assistants can lessen reliance on family members or caregivers (See

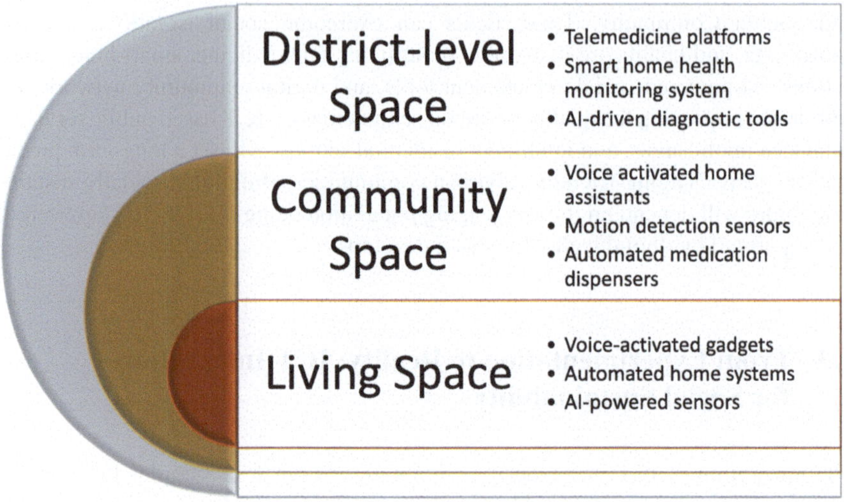

Fig. 5.3 Summary of main suggested ICT-driven interventions to enhance social sustainability of residential compounds at three levels of district-level space, community space, and living space. *Source* The Author's Own

Fig. 5.3). Before expanding based on user feedback and performance assessments, these smart systems can be first tested in a limited number of homes throughout the four communities.

5.3.1 Community-Centered ICT Initiatives for Social Engagement

Active social engagement and mental health are just as important to social sustainability as physical health. Due to the migration of younger family members, especially in Liuxiang Town and Shuijiao Village, many elderly people in suburban Sanya experience social isolation. Community-centered ICT initiatives that emphasize online education, digital social engagement, and intergenerational connection programs can be put into place to combat this.

The launch of regional social networking sites designed specifically for senior citizens, akin to Beijing's "Smart Elderly Care Cloud Platform," is one promising strategy. These platforms can be made to offer seniors virtual cultural and recreational activities, digital literacy training, and connections to local support groups. Regardless of their temporary status, these platforms can help peer-to-peer interaction and help seasonal retirees in Huihui Community, who frequently lack social ties, form new social circles. To guarantee that senior citizens continue to participate in local governance, cultural conversations, and community decision-making, live-streamed community events and interactive forums can be implemented.

ICT-based skill-sharing platforms can empower senior citizens in Jiucai Village, where financial constraints limit access to recreational opportunities, by enabling them to instruct younger generations in traditional crafts, cooking methods, or agricultural knowledge through online workshops. This promotes a sense of purpose and involvement among senior citizens in the community while preserving local heritage through an intergenerational knowledge exchange.

5.3.2 Case Studies of ICT Adoption in Similar Urban Contexts

If we look at similar urban and suburban contexts where technology-driven senior care solutions have been successfully implemented, we can better understand how effective ICT integration is in suburban Sanya. The Shanghai Smart Senior Community Project is one such instance, which implemented AI-powered virtual assistants, intelligent monitoring systems, and a senior-friendly online marketplace in several suburban districts. The project proved the feasibility of smart home interventions and digital healthcare services in aging communities by dramatically increasing senior independence and healthcare accessibility.

A pilot project in Shenzhen's Yantian District combined 5G-powered senior remote medical services, offering video consultations with experts and real-time health monitoring [29]. This project improved early diagnosis rates for chronic illnesses in older patients and decreased hospital visits by 30%. These findings demonstrate the potential influence of ICT-based healthcare solutions in Sanya's suburbs, especially in isolated locations with comparable constraints like Jiucai and Shuijiao Villages.

Japan's Digital Garden City Nation Vision, which prioritizes technology-driven social inclusion for senior citizens in suburban and rural areas, is a pertinent example from around the world [26]. To make sure that senior citizens stayed involved in society, the project implemented VR-based mental health programs, AI-powered senior companionship robots, and intelligent transportation solutions [23]. An analogous model could be implemented in Liuxiang Town, where AI-powered digital companions could help senior citizens by offering emotional support, reminding them of everyday chores, and enabling video calls with relatives who have moved away.

ICT integration in suburban Sanya is a transformative endeavour that adheres to social sustainability principles rather than merely being an experimental process. These suburban communities can close the gap between conventional elderly care models and contemporary technological solutions by implementing AI-driven elderly care systems, digital community networks, telemedicine platforms, and smart home technologies. Long-term digital strategies will be shaped in part by the insights gathered from these pilot projects, guaranteeing that senior citizens in Liuxiang

Town, Jiucai Village, Shuijiao Village, and Huihui Community enjoy better quality of life, increased social interaction, and increased independence.

5.4 Challenges and Opportunities in ICT-Driven Elderly Empowerment

ICT integration in senior living communities in Liuxiang Town, Jiucai Village, Shuijiao Village, and Huihui Community offers both enormous obstacles and game-changing possibilities. Although ICT-driven solutions have the potential to improve social connectivity, healthcare access, and general quality of life, adoption is hampered by the demographic and socioeconomic realities of suburban Sanya. In order to effectively use ICT for social sustainability, it is necessary to address issues of governance, policy limitations, digital literacy gaps, and affordability while simultaneously optimizing the potential for innovation, inclusive digital participation, and community-driven smart aging projects.

The elderly population's low level of digital literacy is one of the main issues in these suburban communities. In Jiucai Village and Shuijiao Village, a large number of seniors—especially those from lower-income families—have little experience with smartphones, internet services, and digital health apps. The adoption of ICT-driven social networking and healthcare tools is hampered by this unfamiliarity. Furthermore, older residents' use of Hainanese dialects and rural linguistic variances makes it difficult to communicate on standard Mandarin-based digital platforms. Another significant problem is affordability, especially for the elderly in Jiucai and Shuijiao Villages, where household incomes are lower than in Sanya's urban areas. The inability of many seniors to purchase smartphones, internet subscriptions, or smart home appliances restricts their access to online financial services, digital community networks, and telemedicine programs. Even though the population of Huihui Community is more economically stable, part-time seasonal retirees may not have permanent access to ICT, which makes it more difficult to maintain digital engagement. Widespread ICT adoption in these regions is still difficult to achieve without focused financial aid initiatives or subsidized internet access.

The support of smart aging initiatives by local governance structures and policies is a crucial component in facilitating ICT-driven elder empowerment. There is currently no comprehensive policy framework in place in Sanya's suburban districts that makes it easier to integrate ICT into senior care services. Although smart healthcare initiatives have been introduced in urban Sanya, these programs have not yet reached rural and suburban areas, where progress is hampered by inconsistent funding allocation and bureaucratic inefficiencies. The promotion of public–private partnerships for ICT implementation is largely dependent on local governance in Liuxiang Town, which has superior infrastructure compared to the other case study areas. However, community-based leadership and local advocacy groups need to be more involved in policy discussions and resource allocation in Jiucai and Shuijiao Villages, where municipal oversight is weaker. ICT initiatives run the risk of continuing to be dispersed and underutilized in the absence of robust governance models to monitor the growth of smart aging communities.

Notwithstanding these obstacles, there are special chances for innovation due to the ICT-driven transformation of the senior communities in suburban Sanya. Pilot projects like AI-driven virtual assistants and smart health monitoring systems could be used as a test site for future expansions into less technologically advanced areas in Huihui Community, where many residents have higher levels of digital literacy. Similar to this, community-led digital literacy workshops could be set up in Liuxiang Town, which has more advanced educational and public service institutions, to teach senior citizens the fundamentals of ICT, online safety, and telehealth applications.

Additionally, by utilizing regional cultural customs and intergenerational educational initiatives, inclusive digital engagement can be encouraged [6, 28]. Younger family members and university volunteers, for instance, could serve as digital mentors in Jiucai Village and Shuijiao Village, helping senior citizens use social media, online banking, and smart home appliances. The digital divide can be closed in a way that is both sustainable and culturally appropriate by integrating ICT education into already-existing social support systems and community customs.

In conclusion, there is a lot of promise for innovation, smart aging policy frameworks, and inclusive community-driven technology adoption [30], even though ICT-driven elderly empowerment in Liuxiang Town, Jiucai Village, Shuijiao Village, and Huihui Community faces obstacles like digital literacy, affordability, and governance gaps. Through focused policy interventions, public–private partnerships, and regional digital education initiatives, ICT can close these gaps and become a potent enabler of social sustainability for Sanya's aging population.

5.5 Conclusion and Future Directions

The integration of ICT into elderly residential communities in Sanya's Tianya District has demonstrated its potential to foster social sustainability by improving healthcare accessibility, reducing social isolation, and enhancing the overall quality of life for aging populations. The sample case studies of Liuxiang Town, Jiucai Village, Shuijiao Village, and Huihui Community highlight the transformative impact of telemedicine, smart home innovations, and digital engagement platforms in ensuring a more inclusive and connected environment for seniors. For long-term advantages and accessibility to be maintained, these advancements must be strengthened by strategic policy initiatives, governance structures, and ongoing research.

Policy interventions must place a high priority on affordability, infrastructure development, and digital literacy initiatives to fully utilize ICT in elderly residential compounds. To implement intergenerational digital learning initiatives, increase broadband access in suburban areas, and subsidize smart home technologies, Tianya District municipal authorities should work with private tech firms, healthcare providers, and neighbourhood organizations. Policies pertaining to data security and user privacy also need to be reinforced to foster trust and promote older users' increased adoption of ICT. In order to guarantee that local expertise and cultural factors influence the creation of ICT solutions suited to the requirements of senior citizens, governance frameworks should encourage community-driven projects.

In order to evaluate the long-term advantages and possible hazards of digital transformation, future research should look into longitudinal studies on the social and psychological effects of ICT use in senior communities. Furthermore, examining the ways in which cutting-edge technologies like robotics-assisted caregiving, AI-driven health monitoring, and virtual reality for cognitive therapy can enhance senior living experiences in suburban Sanya may open the door to creative policy frameworks. In order to identify best practices that can be applied to other aging suburban areas in China and beyond, future research should also examine how scalable these ICT-driven initiatives are.

Lastly, a comprehensive, multi-stakeholder strategy that strikes a balance between policy-driven governance, social sustainability objectives, and technological advancements is needed for the successful integration of ICT into senior living communities. ICT can be a potent enabler of smart aging communities by tackling current issues and seizing new opportunities. This will guarantee that senior citizens living in Sanya's suburbs—and other comparable urban settings—enjoy greater social inclusion, a better quality of life, and general well-being in the digital age.

References

1. Barron P, Cord L, Cuesta J, Espinoza S, Larson G, Woolcock M (2023) Social sustainability in development: meeting the challenges of the 21st century. Washington D.C.: World Bank Publications
2. Bruins MJ, Van Dael P, Eggersdorfer M (2019) The role of nutrients in reducing the risk for noncommunicable diseases during aging. Nutrients11(1):85
3. Chan M, Estève D, Escriba C, Campo E (2008) A review of smart homes—present state and future challenges. Comput Methods Programs Biomed 91(1):55–81
4. Chanda A (2024) AI-driven IoT solutions for managing healthcare data in smart cities. Annals of Healthcare Systems Engineering 1(1):19–28
5. Chen S, Niu M, Ngai CSB (2024) What is the next step of ICT development? the changes of ICT use in promoting elderly healthcare access: a systematic literature review. Heliyon.
6. Cheng H, Lyu K, Li J, Shiu H (2021) Bridging the digital divide for rural older adults by family intergenerational learning: a classroom case in a rural primary school in China. Int J Environ Res Public Health 19(1):371
7. Cheshmehzangi A (2022) ICT, cities, and reaching positive peace. Springer, Singapore
8. Cheshmehzangi A (2022b) The application of ICT and smart technologies in cities and communities: An overview. ICT, cities, and reaching positive peace. Singapore: Springer, 1–16.
9. Cheshmehzangi A (2022c) ICT for high levels of human capital in cities: A shift towards E-government and knowledge-driven economies. In: ICT, cities, and reaching positive peace. Singapore: Springer Nature Singapore, pp 139–157
10. Cheshmehzangi A, Dawodu A, Sharifi A (2021) Sustainable urbanism in China. Routledge, New York and London
11. Cheshmehzangi A, Tang T (2024) China's urbanization progress and urban change: sustainability cannot be divorced from China's future development. In: 30 Years of Urban Change in China's 10 Core Cities. Springer, Singapore, pp 1–25

12. Cheshmehzangi A, Zou T, Su Z, Dawodu A, Ardakani SP, Sedrez M, Tang T (2024) Growing digital divide for expatriates population in China: a user perspective analysis during the COVID-19 Pandemic. SAGE Open 14(4):21582440241274612
13. Cheshmehzangi A, Zou T, Zhang Y, Chen H, Su Z, Dawodu A, Xie L (2022) Commentary: reflecting on the neglected digital divide barriers of telemedicine during COVID-19. Front Public Health 10:915401
14. Demiris G, Hensel BK (2008) Technologies for an aging society: a systematic review of "smart home" applications. Yearb Med Inform 17(01):33–40
15. Fang EF, Xie C, Schenkel JA, Wu C, Long Q, Cui H, Aman Y, Frank J, Liao J, Zou H, Wang NY (2020) A research agenda for ageing in China in the 21st century: Focusing on basic and translational research, long-term care, policy and social networks. Ageing Res Rev 64:101174
16. Freestone R, Favaro P (2022) The social sustainability of smart cities: A conceptual framework. City Cult Soc 29:100460
17. Gyasi RM, Phillips DR (2020) Aging and the rising burden of noncommunicable diseases in sub-Saharan Africa and other low-and middle-income countries: a call for holistic action. Gerontologist 60(5):806–811
18. Huiqin G (2018) Analysis of Influencing Factors of Migratory Tourists Behavior in Sanya. In: Proceedings of 2018 1st International Conference on Education, Art, Management and Social Sciences (EAMSS 2018), pp 68–73, Available online from: https://www.clausiuspress.com/conferences/LNEMSS/EAMSS%202018/A0017.pdf
19. Iuga IC, Nerişanu RA, Iuga H (2024) The impact of healthcare system quality and economic factors on the older adult population: a health economics perspective. Front Public Health 12:1454699
20. Jo E, Epstein DA, Jung H, Kim YH (2023, April) Understanding the benefits and challenges of deploying conversational AI leveraging large language models for public health intervention. In: Proceedings of the 2023 CHI Conference on Human Factors in Computing Systems, pp 1–16
21. Kou L, Xu H, Hannam K (2017) Understanding seasonal mobilities, health and wellbeing to Sanya, China. Soc Sci Med 177:87–99
22. Kou L, Xu H, Kwan MP (2018) Seasonal mobility and well-being of older people: The case of 'Snowbirds' to Sanya, China. Health Place 54:155–163
23. Le DN, Van Le C, Tromp JG, Nguyen, GN (eds) (2018) Emerging technologies for health and medicine: virtual reality, augmented reality, artificial intelligence, internet of things, robotics, industry 4.0
24. Majumder S, Aghayi E, Noferesti M, Memarzadeh-Tehran H, Mondal T, Pang Z, Deen MJ (2017) Smart homes for elderly healthcare—recent advances and research challenges. Sensors 17(11):2496
25. McTear M, Ashurkina M (2024) A new Era in conversational AI. Transforming conversational ai: exploring the power of large language models in interactive conversational agents. Apress, Berkeley, CA, pp 1–16
26. Nath N, Nitanai R, Manabe R, Murayama A (2023) A global-scale review of smart city practice and research focusing on residential neighbourhoods. Habitat Int 142:102963
27. National Bureau of Statistics of China (2021) Main data of the seventh national population census. Available from: https://www.stats.gov.cn/english/PressRelease/202105/t20210510_1817185.html
28. Patrício MR, Osório A (2016) Intergenerational learning with ICT: a case study. Studia paedagogica 21(2):83–99
29. Ren JY, Lei YM, Lei BS, Peng YX, Pan XF, Ye HR, Cui XW (2023) The feasibility and satisfaction study of 5G-based robotic teleultrasound diagnostic system in health check-ups. Front Public Health 11:1149964
30. Wang H, Coyte PC, Shi W, Zong X, Zhong R (2023) Social governance and sustainable development in elderly services: innovative models, strategies, and stakeholder perspectives. Sustainability 15(21):15414

31. Wells M (2010) Resilience in older adults living in rural, suburban, and urban areas. Online J Rural Nurs Health Care 10(2):45–54
32. WHO (World Health Organization) (2021) Guidance on community mental health services: promoting person-centred and rights-based approaches. World Health Organization. Available online from: https://iris.who.int/bitstream/handle/10665/341648/9789240025707-eng.pdf?sequence=1%20pp

Chapter 6
Innovation and Socio-Economic Sustainability: ICT and Social Media for Place Promotion

6.1 Introduction: The Role of ICT and Social Media in Place Promotion

The rapid evolution of information and communication technology (ICT) and the rise of social media platforms have transformed how cities and regions promote themselves in the global digital landscape [14, 23, 35]. Place promotion has moved toward dynamic, real-time, and interactive digital engagement from its traditional reliance on physical branding, travel agencies, and word-of-mouth marketing [9, 42]. In this regard, social media and ICT are effective instruments for promoting travel destinations [21, 34], improving urban experiences, and promoting socioeconomic sustainability. Sanya, a well-known coastal city in southern Hainan Island, China, is a prime example of how digital platforms influence place identity, draw tourists, and boost the economy [38]. Sanya's local economy gains from more tourists, a wider range of business opportunities, and improved cultural visibility when ICT-driven strategies are used.

The tactics used to improve a city's or region's reputation and increase its appeal for long-term residents, business investments, and tourists are collectively referred to as place promotion [32]. Social media sites like WeChat, Douyin (TikTok China), Xiaohongshu (RedNote), and Weibo have completely changed how cities market themselves in the digital age [43]. Digital promotion, in contrast to traditional forms of advertising, enables user-generated content, interactive storytelling, and real-time engagement—all of which together influence public opinion. Cities like Sanya can create an immersive and captivating story through influencer collaborations, live-streamed tourism experiences, and AI-powered marketing campaigns [45], increasing the region's appeal to tourists and investors.

As a continuation of place promotion, digital placemaking makes use of ICT advancements to improve urban areas through augmented reality, interactive media

A. Cheshmehzangi, *Urban Innovation and Sustainability*, Urban Sustainability,
https://doi.org/10.1007/978-981-96-6370-5_6

installations, and data-driven design [16, 33, 40]. Cities can enhance the visitor experience and resident well-being by incorporating mobile-based navigation tools, AI-powered customer service, and smart city infrastructure. For instance, Sanya's urban environment is made more effective and interesting by tourism apps that provide real-time information on the weather, crowd density, and local events. By boosting accessibility and foot traffic, these digital interventions not only improve tourism's economic viability but also help local businesses.

6.1.1 Understanding Place Promotion in the Digital Age

This section provides a theoretical foundation for place promotion, highlighting how cities leverage ICT and social media to enhance their global image, attract investment, and promote tourism. It explores the intersection of placemaking, branding, and digital media strategies in urban sustainability. A key tactic for cities looking to draw in visitors, investors, and new citizens has long been place promotion. This procedure has historically depended on international expos, television campaigns, and print advertising [16]. However, the way cities market themselves has drastically changed as a result of the digital transformation of urban branding. These days, cities can improve their visibility, encourage engagement, and present a powerful and dynamic image by utilizing interactive, data-driven, and real-time tools that are made possible by social media platforms and ICT [18]. Placemaking, branding, and digital media tactics come together in this changing environment to form a potent framework for urban sustainability.

The strategic dissemination of a city's identity, culture, and economic prospects to outside audiences is the fundamental definition of place promotion. Cities can now instantly reach international markets thanks to ICT and digital platforms, enabling a more sophisticated and flexible approach to urban branding [46]. ICT-driven social media platforms are now crucial in determining how cities like Sanya are portrayed. By offering a two-way communication channel, these platforms facilitate public involvement in the perception, marketing, and use of places in addition to promotional outreach (See Fig. 6.1).

6.1.1.1 Digital Placemaking and Smart Urban Branding

The process of transforming urban areas to reflect cultural and economic vibrancy is known as "placemaking" [5, 8, 13, 19, 41], and it is a crucial component of place promotion and urban sustainability. Placemaking has historically involved tangible interventions like public areas, cultural districts, and architectural landmarks [22]. However, placemaking has moved into the digital sphere with the introduction of ICT, combining interactive media and smart urban branding to completely reimagine how people interact and experience cities.

Fig. 6.1 The common tool and device alongside beach areas in Sanya for promoting activities, scenery, and tourism-related businesses. *Source* The Author's Own

Data analytics, AI-powered personalization, and digital storytelling are all used in smart urban branding to give a city an engaging and immersive identity [11]. To appeal to particular visitor demographics, Sanya, for example, uses real-time digital campaigns that highlight local culinary traditions, eco-tourism experiences, and seasonal festivals. Travelers are more likely to interact with local companies and cultural landmarks when they receive personalized recommendations thanks to geo-targeted advertising and AI-driven content curation. This highly customized branding strategy promotes sustainable business practices and enhances Sanya's standing as a top travel destination.

6.1.1.2 The Role of ICT and Social Media in Place Promotion

ICT is essential to contemporary place promotion tactics, especially when it comes to the application of immersive technologies, social media engagement, and big data analytics. Virtual reality (VR) and augmented reality (AR) applications are now used by cities to provide virtual tourism experiences [16], enabling prospective tourists to explore landmarks and destinations before actually visiting. The tourist experience is greatly improved by AI-powered chatbots on platforms such as WeChat [30],

Fig. 6.2 An example of beach side social media activities with social media influencers and users making live coverage of holiday destination in and around Sanya. *Source* The Author's Own

which offer real-time assistance, multilingual support, personalized itineraries, and live updates on local events.

By encouraging user-generated content (UGC), which raises credibility and encourages natural engagement, social media has an impact on place promotion that goes beyond traditional marketing [1, 26, 27, 37]. Cities must create online communities where visitors, influencers, and locals actively participate in the city's developing brand narrative rather than depending only on official campaigns. For instance, Sanya capitalizes on the viral potential of live-streamed tourism experiences, influencer partnerships, and travel vlogs, all of which help to influence the city's image in both local and foreign markets. Live engagement sessions, interactive storytelling, and hashtag campaigns have developed into effective strategies for drawing in tourists from around the world and maintaining urban tourism (See Figs. 6.2, 6.3, 6.4, 6.5, 6.6, and 6.7).

6.1.1.3 Place Promotion and Socio-Economic Sustainability

In addition to marketing, digital place promotion is essential for socioeconomic sustainability because it promotes local companies, cultural representation, and heritage preservation. Smaller, less well-known businesses can increase their visibility in cutthroat travel and investment markets with the help of ICT-enabled promotion. By enabling direct-to-consumer transactions, digital platforms [3, 28, 31] enable

Fig. 6.3 A livestreaming example at the beachside to promote tourism, products, holiday packages, etc. *Source* The Author's Own

Fig. 6.4 Beach side social media or digital platforms to promote tourism, products, holiday packages, etc. *Source* The Author's Own

Fig. 6.5 Those involved in social media or digital platform continue operating till late at nights alongside beaches and public areas in Sanya. *Source* The Author's Own

Fig. 6.6 A group of social media influencers and social media users working on livestreaming sessions and platforms for their online audiences. *Source* The Author's Own

Fig. 6.7 And those who do not comply with policies and regulations are escorted out of non-designated zones by the beach security staff. *Source* The Author's Own

small hospitality enterprises, eco-tourist operators, and local artisans to get around conventional gatekeepers and reach a larger audience.

Additionally, by incorporating ICT into place promotion, urban sustainability objectives are in line with economic growth, resulting in less exploitative and more inclusive tourism [7]. In order to promote equitable growth across urban and suburban areas, cities can prevent crowding in popular districts and promote the exploration of underutilized spaces by employing data analytics to distribute tourist flows [15]. This could entail using digital storytelling in Sanya to promote eco-friendly lodging,

Fig. 6.8 The success to rejuvenating the urbanity of some local places is due to the growing social platforms and their popularity among the locals and tourists across many places, such as Houhai Village in eastern coastal parts of Sanya. *Source* The Author's Own

community-based tourism, and rural heritage sites—all of which would strengthen sustainability and promote economic diversification.

Place promotion has changed in the digital age, becoming a more dynamic, data-driven, and participatory process [10, 24]. Social media and ICT have emerged as crucial instruments for placemaking, urban branding, and sustainable economic growth, enabling cities like Sanya to forge a strong and recognizable identity on a global scale [29]. Sanya can maintain the competitiveness, inclusivity, and environmental consciousness of its tourism industry while promoting long-term socioeconomic sustainability by strategically utilizing digital technologies [39]. Place promotion will depend more and more on innovation-led tactics that strike a balance between digital engagement and sustainable urban growth as technology advances, influencing the future of cities in a globalized world.

Fig. 6.9 Now Houhai Village has become a place with many nodes that visitors could recognise from social media platforms and online marketing platforms. *Source* The Author's Own

6.1.2 Socio-Economic Sustainability and Digital Transformation in Sanya

Focusing on Sanya's socio-economic landscape, this sub-section examines tourism-driven economic models, urban development trends, and the impact of digital transformation on local businesses [17]. It assesses how social media platforms contribute to sustainable economic growth and inclusive urban branding.

Sanya has long been regarded as one of China's top tropical travel destinations. The city's economy depends heavily on tourism because of its immaculate beaches, opulent resorts, and rich cultural legacy. However, Sanya faces both opportunities and challenges in promoting socio-economic sustainability through digital transformation as global travel patterns change and digital technologies reshape urban economies [29]. This section examines how the city's economy is changing, how

ICT is being incorporated into neighbourhood businesses, and how social media is helping to foster inclusive urban branding and sustainable economic growth.

Sanya's economy is largely reliant on tourism, which also plays a major role in urban growth and local employment. Every year, the city welcomes millions of tourists from both domestic and foreign countries, which fuels the growth of the retail, service, and hospitality sectors. Large-scale urban renewal initiatives, smart transportation networks, and high-speed rail connections are just a few of the significant infrastructure advancements fuelled by this expansion. Over-reliance on tourism [2], however, creates vulnerabilities, especially during recessions or major world crises like the COVID-19 pandemic, which brought attention to the dangers of a monolithic economic structure. Sanya has adopted digital transformation as a crucial tactic to improve its economic sustainability and create a more robust and diverse economy. Local businesses have been able to grow beyond the seasonal limitations of tourism thanks to the integration of ICT and e-commerce solutions in a variety of sectors. Virtual reality (VR) experiences, AI-powered customer support, and smart tourism applications have completely changed how tourists engage with the city [36]. These developments allow for year-round interaction with potential tourists via digital platforms, which not only improves the tourism experience but also opens up new revenue streams for nearby companies.

Traditional markets, local entrepreneurs, and small and medium-sized businesses (SMEs) in Sanya have all been significantly impacted by the adoption of digital technologies. For local vendors, social media platforms have become vital marketing tools because they enable them to connect with a larger audience, highlight distinctive cultural experiences, and increase consumer interaction. For instance, live-streaming commerce has made it possible for regional companies, such as seafood markets and upscale handicraft stores, to display their goods in real time, drawing in both tourists and internet buyers across the country. Additionally, geo-targeted advertising and AI-powered recommendation systems have made it simpler for companies to reach particular demographics, increasing sales and visibility. Visitors can now more easily support local businesses thanks to the further simplification of transactions brought about by digital payment platforms like Alipay and WeChat Pay. Digital transformation guarantees that Sanya's economy stays flexible, inclusive, and prepared for the future as it maintains its position as a major international tourism destination.

6.2 Social Media and Digital Placemaking in Sanya

By enabling cities like Sanya to create a vibrant and interactive brand identity, social media has transformed urban branding and economic sustainability. Digital platforms enable user-generated content (UGC), influencer partnerships, and interactive engagement, in contrast to traditional promotional campaigns that depended on centralized messaging. This results in a more genuine and natural city image.

For example, short-form travel vlogs and hashtag campaigns have greatly expanded Sanya's online visibility and drawn in a new generation of tech-savvy tourists. Both locals and tourists share their experiences, opinions, and suggestions, adding to the city's online story. By encouraging genuineness, inclusivity, and community involvement, this decentralized branding strategy supports the city's long-term socioeconomic sustainability objectives.

Social media has also developed into a forum for information exchange and cross-cultural interaction, which enables Sanya to promote lesser-known sights, eco-tourist projects, and locally led experiences. In order to establish a sustainable and socially inclusive urban brand, the city uses digital storytelling to promote local heritage, environmental conservation initiatives, and responsible tourism. Sanya's socioeconomic sustainability as it embraces digital transformation hinges on striking a balance between inclusive growth strategies and technological innovation. Social media marketing and ICT-driven solutions not only promote economic resilience but also change the way the city interacts with its citizens, businesses, and tourists. In order to ensure long-term sustainability and protect its distinctive cultural and environmental assets, Sanya is transforming from a traditional tourism-dependent economy to a more diversified and adaptable urban ecosystem by utilizing digital tools.

6.2.1 Placemaking Through Digital Narratives: From Local Identity to Global Visibility

In the digital age, placemaking encompasses more than just physical infrastructure and urban planning; it is also being influenced by digital narratives that are redefining how people view, interact with, and experience places. The emergence of social media platforms and ICT-driven innovations in Sanya has had a big impact on placemaking tactics, turning the city's local identity into a well-known travel destination. Sanya has been able to improve inclusive urban branding, foster cultural identity, and change its socioeconomic landscape by utilizing platforms like WeChat, Douyin, Xiaohongshu, and Weibo. Influencer marketing, user-generated content (UGC), and digital storytelling have emerged as crucial instruments for promoting community involvement, drawing in investment, and promoting economic sustainability.

In the past, Sanya's natural beauty, tourism industry, and cultural legacy have all influenced the city's identity. However, conventional urban planning and marketing strategies have frequently fallen short of capturing the richness of local experiences or incorporating the community into the narrative of the city. An approach to city promotion that is more interactive, participatory, and data-driven has been made possible by the incorporation of ICT into placemaking strategies. Small companies, regional craftspeople, and cultural institutions can now present their goods, services, and legacy to a worldwide audience through digital platforms. For example, Xiaohongshu (RedNote) has developed into a potent platform for community-driven storytelling, where visitors and locals share their experiences, offer advice, and draw attention to lesser-known facets of Sanya culture. By contributing to a more varied and genuine portrayal of the city, these natural narratives help it differentiate itself from popular commercial tourism narratives.

Additionally, location-based marketing and AI-powered data analytics are used in ICT-driven branding strategies to customize promotional activities and guarantee that various facets of Sanya's identity are seen by the intended audience. For instance, ecotourism projects and cultural heritage sites can draw domestic tourists looking for distinctive, sustainable experiences, while luxury resorts and upscale wellness retreats can advertise to tourists from abroad.

6.2.2 User-Generated Content and Socio-Economic Transformation

User-generated content's (UGC) ability to transform economic opportunities is one of the most important ICT interventions in placemaking. Unlike traditional advertisements, UGC builds trust, relatability, and engagement, making it a valuable driver of economic sustainability. Local businesses can reach a wider audience without depending on traditional advertising thanks to UGC, which is a low-cost, high-impact marketing strategy. For instance, a street food vendor in Sanya's markets may become well-known across the country if a viral Douyin video highlights the distinctiveness of their food. Smaller companies can now compete in the online market thanks to the democratization of visibility, which closes the gap between locally owned businesses and corporately owned tourist destinations.

Additionally, the smooth integration of e-commerce has been made possible by the use of blockchain technology and digital payment systems on platforms such as WeChat. Businesses can now maintain economic activity outside of seasonal tourism cycles by switching from offline sales to online storefronts. Long-term socioeconomic resilience has been fostered by these ICT-driven interventions, which have resulted in a more diverse and inclusive urban economy.

6.3 Socio-Economic Sustainability and Digital Transformation in Sanya

Socio-economic sustainability, which prioritizes long-term economic viability, social inclusivity, and cultural preservation, is intrinsically linked to the use of ICT and social media in place promotion. Utilizing digital tools guarantees a consistent flow of tourists to a city like Sanya, supporting local companies, the hospitality industry, and the artisanal sectors. Social media platforms support fair economic opportunities by giving independent vendors and small businesses online visibility. E-commerce integration in digital tourism platforms promotes economic diversification beyond seasonal tourism peaks by giving local producers, fishermen, and artisans access to larger markets.

Furthermore, the distribution of economic benefits across multiple sectors is guaranteed by AI-driven marketing tools and smart tourism [20]. By encouraging tourists to explore lesser-known locations, personalized digital recommendations based on big data analytics and AI algorithms help reduce over-tourism in busy locations and support sustainable urban mobility [6]. Digital marketing campaigns that showcase ecotourism projects, like the conservation efforts for Hainan's coral reefs or the tropical forest reserves of Sanya, also encourage eco-friendly travel habits, bringing sustainability and economic growth into line [12]. A crucial component of ICT-driven place promotion is social inclusivity, which goes beyond economic sustainability. By enabling local communities to take part in defining their city's identity, digital storytelling makes sure that indigenous narratives and cultural heritage are not overshadowed by commercial branding initiatives. Incorporating local voices, customs, and histories into Sanya's marketing strategies is made possible by social media platforms, which promote a more inclusive and representative urban identity.

Social media and ICT have a role in place promotion that goes far beyond straightforward marketing [4, 25], they are a revolutionary force that is redefining urban sustainability, social inclusivity, and economic growth. Because it allows for real-time engagement, intelligent tourism infrastructure, and fair economic opportunities for local communities, digital innovation has become a key component of Sanya's tourism development. Sanya's urban sustainability initiatives are reinforced by utilizing ICT and digital platforms, guaranteeing inclusive, resilient, and forward-looking growth. As digital technologies advance, they will continue to play a crucial role in urban development in coastal cities like Sanya by influencing socioeconomic sustainability through place promotion.

6.3.1 Influencer Marketing and Cultural Representation in Sanya

Influencer marketing has become a vital part of Sanya's digital placemaking strategy, helping to redefine the city's image and increase its visibility internationally. Influencers produce relatable, experiential content that appeals to niche audiences in contrast to traditional celebrity endorsements. Influencers influence how prospective tourists view, explore, and interact with the city, whether through immersive storytelling, livestreamed shopping, or travel vlogs.

In order to create multifaceted narratives that appeal to various market segments, Weibo and Douyin influencers, for example, regularly document Sanya's changing lifestyle trends, eco-tourism initiatives, and luxury hospitality experiences. While some influencers highlight adventure tourism by showcasing hiking trails, diving locations, and local festivals, others place more emphasis on wellness tourism by highlighting Sanya's spa retreats and traditional Chinese medicine (TCM) experiences. Influencers are essential to cultural representation beyond tourism, making sure that Sanya's traditional markets, indigenous heritage, and local craftsmanship are all integrated into the city's online persona. In order to support community-led conservation efforts, sustainable tourism practices, and responsible travel behaviours, government initiatives have also partnered with digital content creators and social media influencers, highlighting the role of ICT in promoting socio-economic sustainability (See Figs. 6.8 and 6.9).

6.3.2 ICT-Driven Innovations for Inclusive Placemaking

Urban development trends have also been impacted by Sanya's use of ICT for place promotion and economic sustainability, which has shifted the emphasis toward data-driven decision-making and smart city applications. Now, authorities can optimize tourism flows, enhance infrastructure development, and guarantee the fair distribution of economic benefits through the use of digital twin technology, geospatial mapping, and AI-enhanced urban planning. Furthermore, virtual reality (VR) and augmented reality (AR) technologies are emerging as interactive placemaking tools that let visitors and locals alike interact with immersive experiences, cultural landmarks, and historical narratives. By appealing to younger, tech-savvy audiences and allowing older residents to take part in heritage storytelling, these digital tools help close the generational divide.

The "Smart Tourism Platform" is a noteworthy ICT initiative that offers real-time updates on popular tourist destinations, hotel sustainability ratings, and AI-generated travel suggestions based on environmental impact analyses. In addition to enhancing visitor experiences, this platform guarantees that decisions about urban planning are in line with sustainability goals, promoting long-term socioeconomic resilience.

In the digital age, placemaking necessitates a multifaceted, innovation-driven strategy, with social media and ICT acting as dynamic instruments for influencing cultural representation, economic sustainability, and place identity. In Sanya, influencer marketing, user-generated content, and digital storytelling have produced an engaging and inclusive brand narrative that promotes urban resilience and local economic development. Sanya continues to use digital transformation for sustainable urban development by combining AI-powered data insights, immersive technologies, and smart city applications. This keeps its place identity visible globally while maintaining local authenticity.

6.4 Innovation in Place Marketing: Smart Tourism and Data-Driven Promotion

Big data, artificial intelligence (AI), and real-time analytics are driving Sanya's place marketing evolution and changing how the city markets itself as a top travel destination. ICT-driven strategies are now essential in determining Sanya's socioeconomic sustainability due to the rising demand for individualized and technologically enhanced tourism experiences. Local governments and companies are now able to make well-informed decisions, maximize resources, and enhance urban branding through focused digital campaigns thanks to the combination of smart tourism initiatives, geolocation-based promotions, and data-driven marketing.

Sanya's smart tourism has been transformed by big data analytics and artificial intelligence (AI), which have made it possible to better understand visitor preferences, behaviours, and mobility patterns. Through social media platforms, online bookings, e-commerce transactions, and mobile applications, data is continuously gathered, enabling stakeholders to manage tourist flows, forecast tourism trends, and create more adaptable promotional strategies. AI-powered solutions examine user-generated content, social media interactions, and search engine queries to provide insights that guide customized marketing campaigns.

The provincial tourism authority of Hainan, for example, has made investments in AI-powered recommendation engines that make tailored travel recommendations based on user preferences, local events, and current weather conditions. Through integration with Xiaohongshu (Little Red Book or now renamed as 'RedNote'), Douyin (TikTok China), and WeChat Mini Programs, this system offers location-based suggestions that boost visitor satisfaction and highlight lesser-known sights. By preventing overtourism in popular areas and promoting smaller companies and up-and-coming locations, AI-driven place marketing promotes a more evenly distributed tourist population throughout the city.

6.4.1 Geolocation-Based Promotions and Digital Engagement

Utilizing real-time location data to present tailored content to prospective visitors, geolocation-based promotions have become a powerful digital marketing tool. Retailers, lodging establishments, and tour operators in Sanya use location-based services to target visitors with special deals, discounts, and customized experiences as they pass through strategic city districts. For instance, hotels and resorts in Dadonghai Beach and Yalong Bay have incorporated interactive promotions based on QR codes that are connected to Xiaohongshu and Weibo. These promotions allow guests to access private events, unlock exclusive content, and take part in digital competitions. By generating interactive experiences that encourage customers to visit nearby businesses, these geolocation-based tactics improve customer engagement.

Additionally, the municipal government of Sanya has installed augmented reality (AR) navigation systems and smart tourism kiosks in popular tourist destinations like Luhuitou Park and Nanshan Cultural Tourism Zone. These digital touchpoints reinforce Sanya's creative approach to urban placemaking by giving visitors up-to-date information on crowd density, local suggestions, and historical context.

6.4.2 Personalized Digital Experiences and Economic Returns

Data-driven marketing makes personalized digital experiences possible, which are essential to bolstering the tourism industry's economic viability in Sanya. Blockchain-supported loyalty programs, intelligent travel assistants, and AI-powered chatbots are revolutionizing consumer interactions, guaranteeing smooth transactions, and improving overall traveller experiences. For example, a lot of companies in Sanya's night markets and duty-free shopping areas use real-time customer segmentation analytics to provide tailored offers through mobile apps and social media advertisements. By making sure that users receive timely and relevant promotions based on their browsing history and spending patterns, this strategy has dramatically raised conversion rates.

Additionally, to manage real-time transportation data, hotel availability, and event schedules, Sanya's "Smart Destination" initiative incorporates cloud computing and IoT technologies. By reducing inefficiencies and fostering sustainable tourism growth, this interconnected digital ecosystem enables local businesses to optimize profits while guaranteeing a seamless visitor experience.

The combination of big data, artificial intelligence, and real-time analytics will be essential to Sanya's continued embrace of digital transformation in order to grow its tourism industry while preserving social and environmental sustainability. Increased visitor spending and inclusive economic growth have already been shown to result from the integration of smart tourism platforms, geolocation-based marketing, and personalized digital engagement. In the future, Sanya's status as a global center for smart tourism will be further cemented with the use of blockchain technology for transparent transactions, AI-driven predictive analytics for demand forecasting, and virtual reality (VR) tourism experiences. In addition to improving its place promotion tactics, Sanya is reaffirming its dedication to digitally led economic resilience and sustainable urban development with these ICT-driven innovations.

6.5 Discussions: Challenges and Opportunities in ICT-Led Place Promotion

The interaction of technology, tourism, and urban sustainability offers both opportunities and challenges as Sanya continues to establish itself as a world-renowned destination through ICT-led place promotion. A number of obstacles stand in the way of the full implementation of a sustainable and inclusive digital placemaking strategy, despite the fact that social media platforms, big data analytics, and AI-driven marketing have improved Sanya's visibility, tourism economy, and local business growth. This section highlights opportunities in eco-friendly tourism campaigns and smart urban planning initiatives that can foster long-term socio-economic resilience, while also critically examining important issues like the digital divide, algorithm-driven visibility issues, and the risks of overtourism.

The digital divide, which shows up in both the accessibility of local businesses and the ability of visitors to use digital tools, is one of the biggest obstacles to Sanya's ICT-led place promotion. The technological infrastructure and digital literacy required to compete with larger tourism businesses that dominate online visibility on social media platforms are lacking in many small and family-run businesses in Sanya's local markets, night bazaars, and suburban areas. Due to this imbalance, smaller businesses find it difficult to use digital marketing tools, which restricts their ability to compete in the digital tourism market.

Another obstacle is the visibility issues caused by algorithms. Certain attractions, companies, or travel influencers frequently receive disproportionate exposure as a result of social media and search engine algorithms that prioritize content based on engagement metrics and sponsored promotions. While lesser-known cultural and ecotourism destinations in Sanya's hinterlands receive little digital attention, this can further concentrate tourism in already well-known areas like Yalong Bay or Dadonghai Beach. Due to the reliance on algorithm-based visibility, promoting tourism now primarily involves spending money on digital marketing rather than accurately portraying the variety of local experiences.

Another urgent concern is the possibility of overtourism, which is made worse by influencer-driven travel trends and viral digital campaigns. Tourism has increased due to ICT-driven place promotion strategies, particularly during peak seasons. This has resulted in social unrest, environmental stress, and unequal economic benefits. Particularly in coastal regions and nature reserves, the surge of visitors concentrated hotspots causes overcrowding, increased living expenses, and ecological degradation. To ensure that visitor traffic is dispersed equally across several locations and uphold sustainable tourism practices, this calls for a more controlled and balanced approach to digital tourism marketing.

Despite these challenges, ICT-driven place promotion offers transformative opportunities for Sanya's socio-economic sustainability. Campaigns for environmentally conscious and responsible travel represent a sizable opportunity. Promoting eco-friendly travel, low-impact travel, and cultural tourism experiences that boost local economies without unduly harming the environment are all made possible by digital platforms. For instance, WeChat Mini Programs, which provide eco-tourism guides, carbon footprint calculators, and digital incentives for passengers selecting low-emission modes of transportation, have been used to promote sustainable travel practices.

AI-powered tourism management and intelligent urban planning are two more exciting fields. The municipal government and tourism organizations in Sanya can better control visitor flows and ease traffic at popular destinations by utilizing big data analytics and Internet of Things-driven monitoring. Digital visitor analytics, AI-driven crowd control techniques, and real-time data dashboards enable dynamic changes in tourism regulations, like introducing timed entry systems or suggesting different routes to travellers. By preventing economic over-reliance on a small number of important destinations and promoting lesser-known local businesses, these innovations aid in the more equitable distribution of tourist activity throughout the city. Additionally, initiatives for cultural branding and digital storytelling offer underrepresented communities and heritage sites new avenues for visibility. Local crafts, minority cultural customs, and hidden treasures in suburban and rural areas have been highlighted thanks in large part to platforms like Xiaohongshu and Douyin. Through the use of interactive digital campaigns, community-driven narratives, and user-generated content, Sanya can broaden its approach to place promotion and guarantee that the financial advantages go beyond conventional tourist destinations.

It will be crucial to strike a balance between digital innovation and sustainable practices as ICT-driven place promotion continues to influence Sanya's urban and tourism development. A more resilient and inclusive tourism ecosystem can be achieved by addressing the digital divide, enhancing small businesses' equitable online visibility, and controlling overtourism through AI-driven urban planning. Long-term socioeconomic sustainability is also possible by using ICT for cultural preservation initiatives, eco-friendly campaigns, and real-time visitor management. Sanya can increase its appeal on a global scale and promote a more sustainable and advantageous tourism model locally by incorporating smart digital placemaking techniques.

6.6 Future Directions: Strengthening Digital Place Promotion for Urban Sustainability

Emerging technologies and data-driven innovations hold the key to the future of digital tourism marketing and urban sustainability, as Sanya continues to push its ICT-led place promotion strategies. In addition to guaranteeing that financial gains are shared fairly, the combination of immersive virtual tourism, blockchain-based digital identity verification, and AI-powered destination branding is poised to completely transform Sanya's interactions with tourists and investors. Strong governance frameworks, well-thought-out policy interventions, and inclusive participation are necessary for the successful adoption of these innovations in order to meet Sanya's long-term socioeconomic sustainability objectives.

Immersion virtual tourism, which uses augmented reality (AR), virtual reality (VR), and mixed reality (MR) to create immersive, interactive, and remote travel experiences, is one of the most promising areas of future digital place promotion. Virtual reality (VR)-powered tourism can ease physical crowding at major attractions while providing digital access to cultural and ecological sites that need environmental protection, as travel preferences around the world shift toward sustainable and tech-enhanced experiences. For example, Sanya can advertise its attractions without overtaxing delicate ecosystems by incorporating 360-degree virtual tours of Jianfengling's tropical rainforests, Nanshan Temple, and Yalong Bay. Virtual reality experiences can also be used as pre-trip planning tools, encouraging travellers to check out lesser-known local locations before their trip, which helps local businesses that aren't on the typical tourist circuit.

Blockchain technology for digital identity verification is another revolutionary development that improves efficiency, security, and trust in online travel transactions. Blockchain-based solutions can give travellers' safe online personas, guaranteeing easy access to lodging, travel, and event reservations while reducing fraud and privacy issues. As Sanya continues to grow as a global tourism hub, this is especially important for cross-border tourists and high-value tourism segments. Local companies and legislators can also provide customized visitor experiences while upholding data sovereignty and security standards by putting in place decentralized identity management systems.

Furthermore, Sanya's place marketing tactics are about to undergo a revolution thanks to AI-driven destination branding [12]. AI can create hyper-personalized marketing campaigns that target particular demographics, interests, and travel preferences by evaluating big data from social media interactions, search engine trends, and analytics on travel behaviour. AI-driven chatbots and recommendation systems, for instance, can create itineraries for tourists in real time, recommending eco-friendly lodging, genuine cultural encounters, and off-the-beaten-path destinations that complement Sanya's sustainable tourism mission. Additionally, AI can assist in spotting new patterns and changes in traveller behaviour, enabling legislators to make proactive modifications to plans for infrastructure development and marketing campaigns.

To maintain the sustainability, equity, and inclusivity of ICT-led place promotion, policy issues must be resolved even though these innovations present exciting opportunities. The creation of a regulatory framework for digital tourism ethics that addresses data privacy, responsible AI use, and equitable digital representation of local businesses is one important policy recommendation. To avoid monopolization by big tourism companies, it is crucial to guarantee that all tourism stakeholders—small businesses, regional craftspeople, and rural communities—have equal access to digital platforms. Policymakers in Sanya should also concentrate on combining digital marketing techniques with intelligent urban planning to strike a balance between long-term urban sustainability and tourism growth. Big data and artificial intelligence (AI) can guide ecotourism regulations, public transportation improvements, and infrastructure expenditures that support Sanya's overarching development objectives. Sanya can increase its worldwide competitiveness while preserving its socioeconomic and environmental resilience by cultivating a tourism ecosystem that is empowered by digital technology.

In the future, Sanya's ability to combine technological innovation with inclusive, community-driven policies will determine how successful it is at promoting places through ICT. The city can establish a global standard for digital placemaking that is both economically successful and socially sustainable by embracing new trends in immersive tourism [44, 47], blockchain security, and AI-driven branding while upholding moral governance and sustainable urban planning.

References

1. Ali A (2024) Cultivating connections: exploring the influence of user-generated content on social and consumer brand engagement
2. Amin R (2024). Over tourism: a critical analysis from the domestic tourist perspective. Doctoral dissertation, University of Dhaka
3. Anand K (2021) E Commerce a way ahead-direct to consumer. Indian JL & Legal Rsch. 1:1
4. Appel G, Grewal L, Hadi R, Stephen AT (2020) The future of social media in marketing. J Acad Mark Sci 48(1):79–95
5. Arefi M (2014) Deconstructing placemaking: Needs, opportunities, and assets. Routledge, New York
6. Banik P, Banerjee A, Wörndl W (2023, June) Understanding user perspectives on sustainability and fairness in tourism recommender systems. In: Adjunct Proceedings of the 31st ACM Conference on User Modeling, Adaptation and Personalization, pp 241–248
7. Bianchi RV, de Man F (2021) Tourism, inclusive growth and decent work: A political economy critique. Justice and tourism. Routledge, New York and London, pp 220–238
8. Blokland T (2009) Celebrating local histories and defining neighbourhood communities: Place-making in a gentrified neighbourhood. Urban studies 46(8):1593–1610
9. Buhalis D, Foerste M (2015) SoCoMo marketing for travel and tourism: Empowering co-creation of value. J Destin Mark Manag 4(3):151–161
10. Bunders DJ, Varró K (2019) Problematizing data-driven urban practices: Insights from five Dutch 'smart cities.' Cities 93:145–152
11. Chen L, Wang Z, Pang H (2024) Collaborative AI enabled place branding: a hybrid-narrative review and prospects for future research. Open J Soc Sci 12(4):536–566
12. Chen Y, Liu Y, Wang C, Chen K (2019) Is hosting tourism events a sustainable way of destination branding: evidences from Sanya. China. *Cogent Social Sciences* 5(1):1692456
13. Cheshmehzangi A (2020) Reflecting on Urban Identities. In: Identity of cities and city of identities. Singapore: Springer, pp 133–178
14. Cheshmehzangi A (2022) The impacts of ICT platforms on public place use in China. Built Environment 48(2):264–279
15. Cheshmehzangi A (2022b) The application of ICT and smart technologies in cities and communities: An overview. In: ICT, cities, and reaching positive peace. Singapore: Springer, pp 1–16
16. Cheshmehzangi A, Ornsby P (2017) Augmented locality: the utilisation of urban screens in public places as new networks of the city. Journal of Urban Regeneration & Renewal 11(2):194–211

17. Cheshmehzangi A, Tang T (2024) China under construction: shaping cities through recent urban transformation. Singapore; Springer Nature
18. Ciuccarelli P, Lupi G, Simeone L (2014) Visualizing the data city: social media as a source of knowledge for urban planning and management. Cham: Springer, Springer Science & Business Media
19. Dupre K (2019) Trends and gaps in place-making in the context of urban development and tourism: 25 years of literature review. J Place Manag Dev 12(1):102–120
20. Florido-Benítez L, del Alcázar B (2024) How artificial intelligence (ai) is powering new tourism marketing and the future agenda for smart tourist destinations. Electronics 13(21):4151
21. Gössling S (2021) Tourism, technology and ICT: a critical review of affordances and concessions. J Sustain Tour 29(5):733–750
22. Hajzeri A, Deliu S, Sanesi G (2024) Placemaking and the dynamic transformation of urban landscapes: a comprehensive exploration. Int J Cult Policy:1–20
23. Hamelink CJ (1997) New information and communication technologies, social development and cultural change, Vol 86. Geneva: United Nations Research Institute for Social Development. Available online from: https://archive.ccrvoices.org/cdn.agilitycms.com/centre-for-communication-rights/Images/Articles/pdf/Hamelink1997.pdf
24. Kaluarachchi Y (2022) Implementing data-driven smart city applications for future cities. Smart Cities 5(2):455–474
25. Kaplan AM (2012) If you love something, let it go mobile: Mobile marketing and mobile social media 4x4. Bus Horiz 55(2):129–139
26. Kim M, Lee M (2017) Brand-related user-generated content on social media: the roles of source and sponsorship. Internet Res 27(5):1085–1103
27. Kim M, Song D (2020) When brand-related UGC induces effectiveness on social media: the role of content sponsorship and content type. In: Electronic Word of Mouth as a Promotional Technique. Routledge, pp 105–124
28. Leimstoll U, Wölfle R (2021) Direct to Consumer (D2C) e-commerce: goals and strategies of brand manufacturers. New trends in business information systems and technology: Digital innovation and digital business transformation, 237–250
29. Li Y, Alencar A (2023) A tale of two cities: digital place-making and elderly Houniao migration in China. J Ethn Migr Stud 49(4):1032–1049
30. Ma X (2025) Exploring the Impact of AI Chatbot Attributes on Consumer Purchase Intentions in TikTok's Social Commerce Ecosystem in China. Cuestiones de Fisioterapia 54(3):1979–1994
31. McKee S, Sands S, Pallant JI, Cohen J (2023) The evolving direct-to-consumer retail model: a review and research agenda. Int J Consum Stud 47(6):2816–2842
32. Morgan N, Pritchard A, Pride R (2011) Tourism places, brands, and reputation management. Destination brands 3:3–19
33. Najafi P (2024) Digital placemaking and healthy ageing: a user-centric approach for empowering senior citizens in inclusive decision-making of future healthy ageing neighbourhoods. Available online from: https://research.tue.nl/files/339854744/20241008_Najafi_hf.pdf
34. Pencarelli T (2020) The digital revolution in the travel and tourism industry. Information technology & tourism 22(3):455–476
35. Pourroostaei Ardakani S, Cheshmehzangi A (2023) Big data for social media analysis during the COVID-19 Pandemic: an emotion analysis based on influences from social networks. In: Big data analytics for smart transport and healthcare systems. Singapore: Springer Nature Singapore, pp 121–139
36. Rane N, Choudhary S, Rane J (2023, October 31) Sustainable tourism development using leading-edge artificial intelligence (AI), blockchain, internet of things (IoT), augmented reality (AR) and virtual reality (VR) technologies. Blockchain, Internet of Things (IoT), Augmented Reality (AR) and Virtual Reality (VR) technologies
37. Skinner HM (2018) Who really creates the place brand? Considering the role of user generated content in creating and communicating a place identity. Communication and Society 31(4):9–24
38. Sun J, Zhang S, Ji M (2020) Revisiting the impacts of tourism from the perspective of social space production: an ethnological study of the Muslim community in Sanya, Hainan Province China. Current Issues in Tourism 23(15):1845–1863

39. Suryani W (2024) Cultural and heritage tourism trends for sustainable tourism. Spec Interes Trends Sustain Tour:1–15
40. Van Ameijde J (2023) Data-driven urban design: conceptual and methodological constructs for people-oriented public spaces. Disruptive Technologies: The Convergence of New Paradigms in Architecture. Springer International Publishing, Cham, pp 59–84
41. Vaughan J, Maund K, Gajendran T, Lloyd J, Smith C, Cohen M (2021) Determining and representing value in creative placemaking. J Place Manag Dev 14(4):430–445
42. Wan F, Li J (2024) Navigating the digital age: city branding in the era of social media and digital transformation. J Knowl Econ: 1–34
43. Wan R, Tong L, Knearem T, Li TJJ, Huang THK, Wu Q (2025) Hashtag re-appropriation for audience control on recommendation-driven social media Xiaohongshu (rednote). arXiv preprint arXiv:2501.18210
44. Yang W, Cheng Y, Xu C, Zhang J (2023) The seasonal migrants spatially affect the park green space accessibility and equity under different travel modes: evidence from Sanya. China. *ISPRS International Journal of Geo-Information* 12(10):423
45. Yang SH, Tsai CF, Leelapattana W, Thongma W (2025) Tourism marketing strategies from price tags to brand value. Journal of Ecohumanism 4(1):3458–3477
46. Yovanof GS, Hazapis GN (2009) An architectural framework and enabling wireless technologies for digital cities & intelligent urban environments. Wireless Pers Commun 49:445–463
47. Zhao Y (2023) Place-makers in the digital age: lifestyle migrants and the emergence of lifestyle-oriented urbanization in China. J Urban Aff: 1–16

Chapter 7
Innovation and Economic Sustainability: ICT and E-Commerce for the Local Markets

7.1 Introduction: The Digital Transformation of Commerce Industry

A more integrated and effective economic ecosystem has been made possible by the swift development of information and communication technology (ICT), which has completely transformed the commerce sector and altered conventional business models [14, 28, 37]. By enabling companies to reach a wider audience, increase operational effectiveness, and improve customer experiences, digital transformation has emerged as a key factor in boosting economic sustainability [43]. In particular, e-commerce has become a significant force behind innovation, opening up new avenues for companies of all sizes to engage with the global economy [15]. ICT and e-commerce integration are essential for influencing the local economy, encouraging entrepreneurship [34, 46], and advancing sustainable urban growth in cities like Sanya.

Sanya has been at the vanguard of ICT-driven economic transformation due to its heavy reliance on tourism, hospitality, and retail. Small and medium-sized businesses (SMEs) and other local businesses can now reach a wider audience beyond their physical storefronts thanks to digital commerce platforms [21]. Platforms such as Taobao, JD.com, Douyin e-commerce, and WeChat Mini Programs have made it possible for smooth transactions, customized marketing campaigns, and effective supply chain management [6, 47]. Particularly during disruptions like the COVID-19 pandemic, when digital marketplaces became crucial for maintaining business continuity, these innovations have facilitated economic resilience [32]. Additionally, digital payment solutions, AI-powered customer analytics, and smart logistics systems have increased productivity and raised customer satisfaction [38], solidifying Sanya's standing as a leader [44] in smart commerce.

The effect of digital transformation on economic sustainability is a crucial aspect of the commerce sector [7]. Scalability for smaller businesses is limited by the

A. Cheshmehzangi, *Urban Innovation and Sustainability*, Urban Sustainability,
https://doi.org/10.1007/978-981-96-6370-5_7

high operating costs and substantial physical infrastructure requirements of traditional commerce systems. Cloud computing, blockchain, and artificial intelligence-powered e-commerce platforms provide scalable, affordable, and ecologically friendly substitutes for traditional retail [36]. Digital commerce helps create a more efficient and environmentally friendly economy by decreasing the need for physical stores, minimizing resource waste, and improving transportation logistics. The adoption of AI-driven inventory management, automated supply chains, and real-time consumer insights has greatly enhanced Sanya's business performance by enabling retailers to precisely meet customer demands [34], anticipate market trends, and cut surplus. Through resource optimization and the encouragement of wise decision-making, these advancements show how ICT promotes economic sustainability.

The digital transformation of commerce promotes inclusive economic participation in addition to business efficiency [10, 33]. Digital marketplaces give small businesses, artisans, and entrepreneurs in rural areas an equal chance to compete with larger companies. Local farmers, seafood vendors, and handicraft makers in Sanya have benefited greatly from this, as they can now sell their goods online to both domestic and foreign consumers. Producers are able to avoid traditional supply chains, cut down on intermediary costs, and build direct relationships with consumers thanks to platforms like Pinduoduo and livestreaming e-commerce [24]. By promoting indigenous goods and regional specialties, this move toward digitally inclusive commerce not only strengthens economic resilience but also strengthens local communities and protects cultural heritage.

7.1.1 The Rise of Smart Retail and ICT-Driven Commerce

The growth of smart retail and experiential commerce is another important facet of ICT-driven commerce in Sanya [18]. The way consumers interact with brands and products has been completely transformed by the combination of augmented reality (AR), virtual reality (VR), and AI-powered recommendation systems (AI [1]). The city's appeal as a technologically sophisticated business center has increased thanks to the availability of immersive virtual shopping, AI-personalized travel itineraries, and interactive in-store experiences for tourists. Furthermore, advancements in digital wallets, blockchain-secured transactions, and cashless payments have simplified financial interactions and made sure that business transactions are safe, quick, and transparent [2, 5]. In addition to altering customer expectations and behaviour, this shift is also changing how companies organize their operations in a world that is becoming more and more digital.

Despite these developments, there are still obstacles in the way of ICT-driven commerce's broad adoption, especially for small enterprises, the elderly, and traditional market vendors. To guarantee fair participation in the digital economy, concerns like cybersecurity risks, algorithm-driven market competition, digital illiteracy, and regulatory limitations must be addressed [19, 20]. To build a more sustainable and inclusive e-commerce ecosystem, Sanya's business leaders and policymakers must prioritize implementing strong cybersecurity measures, encouraging digital upskilling initiatives, and closing the digital divide. Furthermore, as digital commerce grows, it is important to carefully manage the environmental impact of e-commerce logistics, data privacy issues, and the risk of market saturation through creative governance frameworks and green digital strategies.

In summary, a paradigm shift in economic sustainability is presented by the digital transformation of the commerce sector, which promotes resilience, inclusivity, and efficiency. ICT and e-commerce advancements in Sanya have given local companies more power, increased economic prospects, and established the city as a leader in astute business development. However, long-term sustainability necessitates ongoing investment in digital infrastructure and education, regulatory adaptation, and strategic planning [17]. Sanya's story offers a useful case study for comprehending how ICT-driven commerce can promote economic growth while guaranteeing environmental and social sustainability as technological breakthroughs continue to transform international trade.

7.1.2 Setting the Scene: Economic and Cultural Significance of Local Markets in the Age of E-commerce

Sanya's economy has always been based on its local markets, which are essential gathering places for farmers, small-scale traders, seafood vendors, and regional craftspeople. These marketplaces play a vital role in the city's identity and cultural fabric in addition to offering the community necessities. As traditional markets adjust to a digitalized world, they encounter both opportunities and challenges due to the explosive growth of e-commerce. In addition to navigating the changes brought about by the e-commerce revolution, this section examines how Sanya's local markets continue to play a crucial role in the local economy and cultural heritage.

Sanya's traditional markets have been vital hubs for regional economic activity for centuries. They are the main location for local farmers, seafood vendors, and small-scale traders to sell their goods directly to customers, fostering a vibrant and decentralized economy. In addition to being places for business, markets such as Dadonghai Market and Sanya Fishermen's Market also serve as hubs for the exchange of regional goods, services, and crafts. Particularly for people and families in Sanya's more rural areas who depend on agriculture and fishing as their main sources of income, these markets have strengthened local livelihoods, promoted community ties, and given them a way to become economically empowered.

Another significant economic element that interacts with these markets is tourism [30, 42]. A major component of Sanya's tourism experience is its local markets, where tourists frequently look for genuine experiences, such as trying freshly caught seafood or buying handicrafts made locally. Markets are regarded as cultural icons that offer visitors a glimpse of the customs and way of life of the area. Local markets' offerings and structure have been greatly influenced by the consumption patterns of tourists. Many choose to visit these markets as a cultural experience and to buy goods that represent the local way of life, as tourists swarm to the island for its beaches, resorts, and natural attractions. The integration of traditional markets into the tourism ecosystem has been pivotal in establishing Sanya as an attractive destination for both domestic and international travellers.

In the age of digital commerce, these markets face many obstacles despite the enormous benefits they offer to nearby communities [29, 39]. More people are choosing online shopping because of its convenience and variety as a result of the quick expansion of e-commerce sites like Taobao, JD.com, and Douyin. Local vendors are finding it more difficult to compete with the pricing power, variety, and reach of larger digital platforms as a result of the shift to online marketplaces [22, 35]. Traditional markets run the risk of losing customers as digital retail platforms become more integrated into consumer lifestyles, especially among younger generations who find online shopping more convenient.

However, traditional markets also have a chance to adjust by using digital tools and technologies to become more competitive as a result of this disruption. One possible remedy is the incorporation of e-commerce into regional markets. Market vendors can reach a larger consumer base by selling their goods online, including both locals and visitors who are not physically in Sanya. For example, Sanya seafood vendors can sell directly to consumers without the need for physical storefronts by showcasing their fresh catches on social media sites like Douyin (TikTok) and livestreaming. By reaching travellers from all over the world who are keen to taste the genuine flavours of Sanya's coastal region, these vendors can close the gap between traditional market offerings and the digital economy.

Furthermore, local market vendors can increase the effectiveness of their delivery and transaction systems by incorporating ICT-driven innovations like digital wallets, geolocation-based promotions, and smart logistics [45]. Vendors may find it easier to handle the challenges of digital competition if they are able to expand their customer base beyond their local area and offer better customer experiences. These vendors can benefit from data analytics by better understanding customer behaviour, tailoring their services to particular client segments, and modifying their offerings in response to market trends.

It is important to acknowledge the cultural significance of local markets and how they support Sanya's socioeconomic sustainability, even though the digital transformation of commerce opens up new opportunities. These marketplaces are centers of culture that represent the island's identity in addition to being places of commerce. These markets showcase the rich cultural legacy that sets Sanya apart from other travel destinations, from the traditional foods offered by regional farmers to the handicrafts created by artisans [11]. If local markets adopt contemporary technologies while maintaining their genuine charm, cultural preservation and digital innovation can coexist.

Even though e-commerce is still changing the face of commerce, Sanya's local markets continue to have economic and cultural significance. These marketplaces are crucial for sustaining regional economies, building community ties, and conserving cultural customs. The local markets in Sanya will probably have to strike a careful balance between the advantages of digital innovation and conventional market practices in the future. These markets can develop into hybrid commercial hubs by implementing ICT-driven strategies, providing both digital and physical experiences that cater to the needs of both domestic and foreign tourists. The ability of these markets to innovate while maintaining the local identity and cultural legacy that have made them a vital component of Sanya's socioeconomic landscape will ultimately determine their success in the era of e-commerce.

7.2 ICT and E-Commerce Adoption in Key Local Markets

The way major local markets in Sanya operate has changed dramatically as a result of the quick development of ICT and e-commerce. These markets, which have historically relied on in-person contacts, are now incorporating digital tools, smartphone apps, and online marketplaces to increase customer satisfaction and business efficiency. In order to reach tech-savvy customers who value the ease of online shopping, vendors at markets like Yazhou Central Market and First Market are progressively implementing e-commerce platforms.

The use of digital payment methods, like Alipay and WeChat Pay, has superseded cash transactions, making purchases easier for both residents and visitors. This is one of the most obvious changes. Customers can also pre-order local produce, specialty items, and fresh seafood through online ordering platforms, which cuts down on wait times and increases market accessibility. Logistics networks, such as AI-driven supply chain management and smart delivery services, further support these digital improvements by making sure that goods are delivered from suppliers to customers in an efficient manner [4]. By adopting ICT-driven solutions, Sanya's local markets are striking a balance between tradition and technological advancement, modernizing their business models and guaranteeing their sustainability in an increasingly digital economy.

7.2.1 Case Studies of Digital Integration in Sanya's Markets

Local commerce has changed as a result of ICT integration into Sanya's traditional markets, which has improved customer engagement and operational efficiency. Yazhou Seafood Market (崖州海鲜市场), Hongqi Market (红旗市场), and First Market (第一市场) are three important markets that are excellent illustrations of how digital technologies are changing market dynamics.

7.2.1.1 First Market: A Digital Hub for Tourists and Locals

As one of Sanya's busiest markets, First Market has rapidly adopted mobile payment systems, live-stream selling, and digital engagement strategies to meet the demands of tech-savvy consumers. Due to vendors' heavy reliance on WeChat Pay and Alipay, cash transactions are all but extinct. Nowadays, a lot of seafood vendors use live-streaming websites like Douyin and Kuaishou to display their fresh seafood, communicate with prospective customers in real time, and arrange home delivery. Furthermore, Xiaohongshu (RedNote) digital promotions and online reviews have increased visibility among domestic travellers looking for regional specialties.

7.2.1.2 Hongqi Market: E-Commerce and Smart Transactions

Hongqi Market, which is well-known for its locally produced goods and fresh produce, has modernized its business environment by implementing smart kiosks and QR code-based transactions. Residents can place large orders at a discount thanks to the participation of numerous vendors in community group-buying programs through WeChat Mini Programs. This system improves affordability, minimizes waste, and streamlines logistics. In order to ensure a more sustainable supply chain, some retailers have also incorporated AI-driven inventory management tools to track stock levels and forecast demand trends.

7.2.1.3 Yazhou Seafood Market: Enhancing Customer Experience with Digital Tools

Yazhou District's top seafood market has adopted a data-driven strategy to boost vendor productivity and customer satisfaction. In order to increase transparency and decrease negotiation time, many seafood vendors use digital price boards that update in real-time based on supply conditions. Furthermore, direct seafood deliveries are made possible by e-commerce platforms such as Meituan and Ele.me, which serve both individual customers and restaurants. In order to streamline transactions and lessen disagreements over pricing and weight accuracy, the market has also tested smart weighing systems that interface with digital receipts.

These case studies show how Sanya's traditional markets are changing due to ICT-driven innovations, which are promoting economic sustainability while maintaining their cultural identity.

7.3 Challenges and Barriers to Digital Transformation

The shift to digital commerce has not been without challenges, despite the fact that ICT-driven interventions have significantly increased Sanya's local markets' efficiency. When it comes to implementing e-commerce and digital tools, all three sample local markets faced different obstacles. These obstacles are caused by differences in digital literacy, limitations in infrastructure, complicated regulations, and opposition from conventional suppliers [40]. Furthermore, economic sustainability is hampered by socioeconomic disparities that still restrict equitable access to digital transformation (DT) (Fig. 7.1).

During our survey work, we noticed that digital literacy is a problem for many vendors, especially the older ones in Yazhou Seafood Market and First Market.

Fig. 7.1 Summary of digital transformation (DT) and its key challenges to economic sustainability

Live-stream marketing tactics, online ordering platforms, and mobile payment platforms are often unfamiliar to traditional business owners who have been in these markets for decades. Older vendors prefer cash transactions and in-person negotiations because they are afraid that digital tools will make their business more difficult, while younger merchants are quick to adopt platforms like Douyin, WeChat Mini Programs, and Xiaohongshu. Additionally, there is a natural resistance to altering business models. Vendors who are used to dealing directly with customers are wary of e-commerce platforms because they think that these channels may make it harder to build personal connections with customers. This hesitancy is especially noticeable in Yazhou Seafood Market, where high-value seafood transactions depend heavily on vendor and customer trust.

Although Sanya has made strides in the development of smart cities, Hongqi Market and Yazhou Seafood Market still have uneven digital infrastructure. Rural and suburban markets struggle with signal instability and limited access to high-speed broadband, while First Market, situated in a popular tourist area, enjoys the advantages of robust internet connectivity and widespread smartphone usage. Particularly in seafood markets where freshness and immediacy are crucial selling points, these restrictions impede live-stream commerce, seamless digital transactions, and real-time price updates. Small vendors are also concerned about the cost of ICT upgrades. Many Hongqi Market vendors are unable to make the upfront investments necessary to install digital payment terminals, AI-driven inventory tools, and smart weighing systems. A fragmented business environment results from this unequal adoption rate, where some vendors continue to operate with cash while others switch to digital systems.

ICT adoption in local markets presents regulatory issues, especially regarding vendor licensing, data protection, and e-commerce taxes. First Market and Hongqi Market are home to a large number of small vendors who work in the informal economy and frequently do not have official business registrations. Vendors who previously operated with little oversight now face administrative challenges to continue digital transactions as e-commerce platforms impose stricter regulations on seller verification and tax compliance. Furthermore, there have been conflicting responses to government-led market digitization initiatives. Some vendors believe that regulations make it more difficult for small-scale traders to compete by favouring larger businesses and internet retailers. Transparency is increased by digital tools, but platform service fees, delivery logistics, and advertising costs on WeChat, Douyin, and Ele.me expose companies to increased operating costs.

Not all vendors and customers gain equally from the increased e-commerce opportunities brought about by First Market and Hongqi Market's digital transformation. Access to smartphones, digital banking, and online marketing training are all impacted by socioeconomic disparities, especially for older and low-income traders. Many vendors are at a disadvantage to tech-savvy rivals and big retailers because they lack the funds to invest in expert digital branding, search engine optimization (SEO), and social media advertising. Additionally, consumers from lower-income backgrounds might not have access to digital payment methods, creating a divide between those who are still dependent on cash transactions and those who are digitally connected. This is particularly noticeable in Hongqi Market, where locals who might not have as much access to the internet as visitors and affluent city dwellers are served by fresh produce vendors (See Fig. 7.2).

There are both opportunities and challenges associated with First Market, Hongqi Market, and Yazhou Seafood Market's digital transformation. ICT interventions improve efficiency, outreach, and transparency, but their smooth adoption is hampered by issues like digital illiteracy, inadequate infrastructure, difficult regulations, and socioeconomic disparities. Targeted digital literacy initiatives, reasonably priced ICT infrastructure, and inclusive policies that guarantee financial stability for all suppliers and customers in Sanya's dynamic market ecosystem are necessary to remove these obstacles.

Fig. 7.2 The locals are preparing to set up one of the main local markets of Sanya before 6pm in the evening every day. *Source* The Author's Own

7.4 The Impact of ICT and E-Commerce on Market Efficiency and Sustainability

Sanya's digital transformation has improved financial inclusion for small businesses, streamlined supply chain logistics, and greatly increased market efficiency in the three local markets that were sampled. Digital inventory management, e-commerce platforms, mobile payment systems, and AI-powered logistics are just a few examples of ICT-driven innovations that have changed conventional market dynamics and promoted economic sustainability. Additionally, by reducing waste, optimizing resource use, and lowering carbon footprints through digitalized trade practices, these technological advancements support environmental sustainability.

Especially in First Market, a significant fresh food hub serving both locals and visitors, the combination of mobile payments and online ordering platforms has completely transformed market efficiency. WeChat Pay, Alipay, and UnionPay QR codes have made digital transactions much faster, enabling vendors to serve more customers every day. Furthermore, pre-orders and delivery services are made possible by e-commerce integration through Ele.me (饿了么) and WeChat Mini Programs, which reduces crowding and boosts operational effectiveness.

Digital inventory management tools have aided small-scale farmers and retailers in Hongqi Market by automating reordering procedures, tracking stock levels, and keeping an eye on price fluctuations. Cloud-based inventory systems, which offer real-time data on sales trends, demand forecasting, and supply chain bottlenecks, have replaced the manual stock checks that vendors once used. Cost efficiency has increased as a result of this shift, which has decreased overstocking and product waste. Digital temperature-monitoring systems make sure that seafood stays in ideal condition before it is consumed by customers at Yazhou Seafood Market, where freshness is essential. IoT-enabled tracking devices are now integrated by vendors, enabling them to keep an eye on the condition of seafood while it is being transported. For both suppliers and customers, this improves market reliability, minimizes waste, and stops spoiling.

In all three markets, the use of AI-driven logistics and big data analytics has greatly increased supply chain efficiency. Real-time demand prediction models are advantageous to First Market because they enable wholesalers to dynamically modify pricing and stock levels in response to seasonal variations and customer buying trends. Now that vendors can source products effectively, inventory spoilage and procurement costs are decreased. Hongqi Market, which primarily caters to local customers, optimizes last-mile logistics by utilizing AI-powered delivery algorithms. To speed up deliveries and guarantee product freshness, small-scale farmers and fresh produce vendors work with intelligent logistics platforms like Meituan (美团) and JD.com's cold chain services. Local farmers also have more negotiating power when blockchain-enabled supply tracking is integrated because they can verify the quality of their products and bargain with wholesalers and retailers for reasonable prices.

By eliminating traditional middlemen and connecting seafood vendors directly with buyers throughout China, AI-driven auction platforms enable Yazhou Seafood

Market to secure higher profit margins. By offering real-time pricing transparency, these digital platforms guarantee a more fair and effective trading environment. Smart logistics networks also guarantee that perishable seafood gets to its destination more quickly, which lowers transportation-related emissions overall and improves supply chain sustainability.

7.4.1 Financial Inclusion and Economic Empowerment Through E-Commerce

ICT adoption has been particularly impactful in enhancing financial inclusion for small and informal market vendors [26]. Many vendors in Yazhou Seafood Market and Hongqi Market have historically only accepted cash, which has restricted their access to credit and opportunities for financial advancement. Small businesses can now access microloans from fintech platforms like Ant Financial and WeBank by using transaction records to create credit histories as a result of the adoption of digital payments.

Increased market reach has also resulted from e-commerce integration, particularly for Yazhou Market seafood vendors who can now sell their goods to consumers across the country by live-streaming their transactions on Douyin and Taobao Live. By enabling vendors to reach beyond their immediate customer base, these digital sales channels have increased revenue and strengthened the economy. The demand for local produce has increased as a result of First Market's collaborations with online delivery services, which is advantageous for local farmers and vendors alike. Small-scale farmers can now reach a wider consumer base by utilizing digital trade, which boosts rural economies and lessens income gaps [13] between urban and rural stakeholders.

7.4.2 Sustainability Through ICT-Enabled Waste Reduction and Green E-Commerce

In Sanya's local markets, ICT-driven innovations have also been instrumental in improving environmental sustainability [16]. First Market and Hongqi Market's smart waste management systems improve waste processing efficiency by separating recyclables from organic waste using AI-powered sorting mechanisms. Customers can make more environmentally friendly purchases when supply chain tracking is done with blockchain technology [9], which guarantees increased transparency in sourcing procedures.

Digital tracking systems at Yazhou Seafood Market reduce food waste by synchronizing supply with real-time demand information [12]. In order to reduce excess seafood stock and prevent needless overfishing, vendors are implementing pre-order models through e-commerce platforms. Additionally, incentive-based digital programs are being used to promote sustainable packaging solutions by rewarding vendors who choose biodegradable packaging over alternatives made of plastic. Using data analytics to control energy use in indoor markets is another example of a sustainable innovation. By tracking electricity consumption and optimizing cooling systems, digital energy-monitoring systems enable vendors in Hongqi Market to lower their overall carbon footprints. AI-powered demand-response energy management, automated cooling systems, and smart lighting all help to reduce operating costs and improve environmental impact.

The digital transformation of First Market, Hongqi Market, and Yazhou Seafood Market has significantly improved market efficiency, financial inclusion, and supply chain transparency. Traditional markets have been modernized while maintaining economic and environmental sustainability through ICT-driven interventions, which range from digital inventory systems and cashless transactions to AI-powered logistics and e-commerce platforms. These innovations are improving economic resilience and creating a more sustainable and inclusive local market ecosystem in Sanya by decreasing waste, streamlining operations, and increasing vendor participation in digital trade (See Figs. 7.3, 7.4, 7.5, and 7.6 for selected examples across Sanya).

Fig. 7.3 Despite the apparent digital divide, markets across Sanya are known to many through social media platforms. *Source* The Author's Own

Fig. 7.4 There is this hidden layer of growing e-commerce in most local markets in Sanya. *Source* The Author's Own

Fig. 7.5 Local markets are fantastic places for ICT-driven and ICT-based opportunities, through payments and trade, to e-commerce and outward businesses that help shape the city's vibrant hubs of commercial use. *Source* The Author's Own

Fig. 7.6 Local Markets have become places where e-payments, e-commerce, and e-businesses operate and thrive in clusters. *Source* The Author's Own

7.5 Policy and Governance: Enabling a Digital Market Economy

Effective policy frameworks, financial incentives, and strategic governance are critical to the successful digital transformation of Sanya's traditional markets [25]. In this changing environment, local and regional governments are essential in promoting ICT adoption, controlling online trade, and guaranteeing economic inclusivity. Sanya's shift to a more connected and technologically advanced marketplace has been greatly influenced by the policies of the Hainan Free Trade Port (HFTP) [3, 8, 23, 41], which place a strong emphasis on the expansion of the digital economy.

The creation of regulatory frameworks that strike a balance between consumer protection and market innovation has been one of the main forces behind policy. To entice more vendors to use digital payment methods and online trading platforms, authorities have simplified business licensing procedures, implemented e-commerce tax incentives, and strengthened data privacy laws. These policies have promoted greater financial inclusion and economic resilience by removing obstacles for small-scale vendors in First Market, Hongqi Market, and Yazhou Seafood Market.

ICT integration in local markets has been further expedited by public–private partnerships (PPPs) [27, 31]. Training programs for digital literacy, logistics optimization, and the adoption of mobile payments have been made possible by partnerships with Alibaba, Tencent, and JD.com. Additional assistance for vendors making the switch to online sales is offered by government-sponsored programs like grants for tech-enabled businesses and low-interest loans for digital transformation. In order to promote sustainable digital market economies and preserve Sanya's cultural and economic legacy, policymakers must concentrate on closing the digital divide going forward by increasing access to affordable internet, strengthening cybersecurity safeguards, and creating AI-powered regulatory frameworks.

7.5.1 Future Prospects: Scaling Digital Innovation for Economic Growth

The future of Sanya's traditional markets, including First Market, Hongqi Market, and Yazhou Seafood Market, lies in scaling ICT-driven innovation to foster long-term economic sustainability. Emerging technologies like blockchain, AI-driven analytics, and smart market infrastructure offer new chances to improve efficiency, transparency, and inclusivity as e-commerce and digital platforms become more integrated into the local market ecosystem. AI-driven analytics, which can improve pricing strategies, demand forecasting, and market operations, is one of the main growth areas. First Market vendors can create more sustainable business models by using real-time data to better manage inventory, modify prices in response to market trends, and cut waste. By providing tailored suggestions and focused promotions, AI can also improve customer engagement, raising customer satisfaction and revenue.

Blockchain technology provides a revolutionary way to increase food safety and supply chain transparency. Blockchain-enabled tracking can monitor sourcing, assure authenticity, and build consumer and vendor trust in markets like Hongqi Market, which are dominated by fresh produce and seafood. Additionally, a decentralized digital ledger can strengthen economic resilience by lowering fraud, streamlining transactions, and fostering a more secure and accountable trading environment.

Future-proofing Sanya's local markets will require investments in smart market infrastructure. Adding IoT sensors, automated payment systems, and intelligent logistics networks to physical spaces can boost customer convenience and operational effectiveness. For example, incorporating digital inventory tracking and automated temperature control for seafood storage at Yazhou Seafood Market can enhance supply chain management and lower spoilage, both of which are in line with environmental sustainability objectives.

Programs for digital literacy must be expanded in order to guarantee that ICT integration stays inclusive. Many vendors who are older or less tech-savvy still have trouble using mobile payment systems and e-commerce platforms. To close the digital divide, training programs, funding for digital adoption, and increased internet accessibility should be given top priority in local government initiatives and public–private partnerships. The advantages of digital transformation may be limited in traditional market settings if inclusive policies are not implemented, increasing the risk of economic marginalization.

Ultimately, expanding ICT-driven innovation in Sanya's market economy necessitates a comprehensive strategy that strikes a balance between equitable and sustainable economic growth and technological advancements. Sanya's markets have the potential to develop into digitally enabled trade hubs that support local livelihoods, protect cultural heritage, and promote long-term economic sustainability if they make strategic investments in AI, blockchain, and smart market infrastructure while maintaining widespread accessibility and regulatory oversight.

References

1. Al Balushi JSG, Al Jabri MIA, Palarimath S, Maran P, Thenmozhi K, Balakumar C (2024, June) Incorporating artificial intelligence powered immersive realities to improve learning using virtual reality (VR) and augmented reality (AR) technology. In: 2024 3rd International Conference on Applied Artificial Intelligence and Computing (ICAAIC). IEEE, pp 760–765
2. Alenizi A, Mishra S, Baihan A (2024) Enhancing secure financial transactions through the synergy of blockchain and artificial intelligence. Ain Shams Engineering Journal 15(6):102733
3. An B (2019) Hainan free trade port development. ADB Briefs, No. 118, Nov 2019, Available online from: https://www.adb.org/sites/default/files/publication/542421/adb-brief-118-hainan-free-trade-port-development.pdf
4. Attah RU, Garba BMP, Gil-Ozoudeh I, Iwuanyanwu O (2024) Enhancing supply chain resilience through artificial intelligence: Analyzing problem-solving approaches in logistics management. International Journal of Management & Entrepreneurship Research 5(12):3248–3265
5. Bhat SA, Huang NF, Sofi IB, Sultan M (2021) Agriculture-food supply chain management based on blockchain and IoT: A narrative on enterprise blockchain interoperability. Agriculture 12(1):40
6. Biagiola A (2022) Social commerce in China: the new frontier of e-commerce in the luxury fashion industry. Available online from: https://unive.unitesi.cineca.it/handle/20.500.14247/12418
7. Bocean CG, Vărzaru AA (2023) EU countries' digital transformation, economic performance, and sustainability analysis. Humanities and Social Sciences Communications 10(1):1–15
8. Cao X, Chongmin W (2023) International Experience and Hainan's Path to Building a Free Trade Port. China Econ Transition= Dangdai Zhongguo Jingji Zhuanxing Yanjiu 6(2):180–195
9. Centobelli P, Cerchione R, Del Vecchio P, Oropallo E, Secundo G (2022) Blockchain technology for bridging trust, traceability and transparency in circular supply chain. Information & Management 59(7):103508
10. Chen CL, Lin YC, Chen WH, Chao CF, Pandia H (2021) Role of government to enhance digital transformation in small service business. Sustainability 13(3):1028
11. Chen J, Chen N, Cai F (2022) Sanya, China: Tropical climate and the rise of a small city. Cities 123:103589
12. Chen T, Lv L, Wang D, Zhang J, Yang Y, Zhao Z, Tao D (2024) Empowering agrifood system with artificial intelligence: A survey of the progress, challenges and opportunities. ACM Comput Surv 57(2):1–37
13. Cheshmehzangi A (2016) China's New-Type Urbanisation Plan (NUP) and the foreseeing challenges for decarbonization of cities: a review. Energy Procedia 104:146–152
14. Cheshmehzangi A (2022a) ICT for high levels of human capital in cities: a shift towards E-government and knowledge-driven economies. In ICT, cities, and reaching positive peace. Singapore: Springer Nature, pp 139–157
15. Cheshmehzangi A 2022b. ICT for a sound business environment in cities: opportunities through digital technologies and digital transformations. In ICT, cities, and reaching positive peace. Singapore: Springer Nature, pp 197–217
16. Cheshmehzangi A (2022c) The role of ICT and smartness for peace-building processes and practices. In: ICT, Cities, and Reaching Positive Peace. Singapore: Springer Nature Singapore, pp 17–46

17. Cheshmehzangi A, Tang T (2022) China's carbon neutrality plan implementation and sustainable development process review. In: China's city cluster development in the race to carbon neutrality. Singapore: Springer Nature Singapore, pp 1–23
18. Cheshmehzangi A, Tang T (2024) China under construction: shaping cities through recent urban transformation. Sinagpore: Springer Nature
19. Cheshmehzangi A, Zou T, Zhang Y, Chen H, Su Z, Dawodu A, Xie L (2022) Commentary: reflecting on the neglected digital divide barriers of telemedicine during COVID-19. Front Public Health 10:915401
20. Cheshmehzangi A, Zou T, Su Z, Tang T (2023) The growing digital divide in education among primary and secondary children during the COVID-19 pandemic: An overview of social exclusion and education equality issues. J Hum Behav Soc Environ 33(3):434–449
21. Delgado-de Miguel JF, Buil-Lopez T, Esteban-Navarro MÁ, García-Madurga MÁ (2019) Proximity trade and urban sustainability: small retailers' expectations towards local online marketplaces. Sustainability 11(24):7199
22. Del Mastio N (2021) Successful strategies adopted by foreign brands to thrive in the Chinese e-commerce Market. Available online from: https://unive.unitesi.cineca.it/handle/20.500.14247/8552
23. Dong Q, Bian Z (2024) Meeting international economic and trade rules: a case study of Hainan free trade port China. World Customs Journal 18(1):81–99
24. Feng X (2024) Control, exploitation and exclusion: experiences of small farmer e-tailers in agricultural e-commerce in China. J Agrar Chang 24(1):e12567
25. Han L (2023) Digital economy empowering the transformation of Hainan's modern service industry. Journal of Social Science Humanities and Literature 6(6):57–64
26. Hannig A, Jansen, S (2010) Financial inclusion and financial stability: current policy issues (No. 259). ADBI Working Paper. Available from: https://www.econstor.eu/handle/10419/53699
27. Liu T, Yang X, Zheng Y (2020) Understanding the evolution of public–private partnerships in Chinese e-government: four stages of development. Asia Pacific Journal of Public Administration 42(4):222–247
28. Ma Y, Rong K, Mangalagiu D, Thornton TF, Zhu D (2018) Co-evolution between urban sustainability and business ecosystem innovation: evidence from the sharing mobility sector in Shanghai. J Clean Prod 188:942–953
29. Malecki EJ (2003) Digital development in rural areas: potentials and pitfalls. J Rural Stud 19(2):201–214
30. Meng L, Liu Y, Wang Y, Li X (2021) A big-data approach for investigating destination image gap in Sanya City: When will the online and the offline goes parted? Regional Sustainability 2(1):98–108
31. Nel D (2020) Allocation of risk in public private partnerships in information and communications technology. International Journal of eBusiness and eGovernment Studies 12(1):17–32
32. Nosike CJ, Ojobor OSN, Nosike CU (2024) Enhancing business resilience: Innovation and adaptation during and after the global pandemic. International Journal of Financial, Accounting, and Management 6(2):217–229
33. Otarinia MA (2024) The impact of E-commerce and foreign investment on economic growth in developing countries. Creat Econ New Bus Manag Approaches, 18–59
34. Qiao S (2024) Discussion on network marketing strategy of pearl breeding industry in Sanya. Academic Journal of Business & Management 6(4):144–152
35. Rahman MD (2024) E-commerce Evolution: a Strategic Analysis of Alibaba's Business Ecosystem. Available at SSRN 5109165
36. Sahoo SK, Goswami SS, Sarkar S, Mitra S (2023) A review of digital transformation and industry 4.0 in supply chain management for small and medium-sized enterprises. Spectrum of engineering and management sciences 1(1):58–72
37. Shi Y, Wei, F (2024) Comparative analysis of digital economy-driven innovation development in China: an international perspective. J Knowl Econ:1–43.
38. Shobhana N (2024) AI-powered supply chains towards greater efficiency. In: Complex AI Dynamics and Interactions in Management. Hershey, Pennsylvania: IGI Global, pp 229–249

39. Stockdale R, Standing C (2004) Benefits and barriers of electronic marketplace participation: an SME perspective. J Enterp Inf Manag 17(4):301–311
40. Thompson KM, Jaeger PT, Taylor NG, Subramaniam M, Bertot JC (2014) Digital literacy and digital inclusion: Information policy and the public library. Rowman & Littlefield, Lanham, Maryland
41. Wang S, Jiang X, Khaskheli MB (2024) The role of technology in the digital economy's sustainable development of Hainan free trade port and genetic testing: cloud computing and digital Law. Sustainability 16(14):6025
42. Wu YF, Xu HG, Lew AA (2015) Consumption-led mobilized urbanism: socio-spatial separation in the second-home city of Sanya. Mobilities 10(1):136–154
43. Wujarso R (2023) Effect of digital transformation on company operational efficiency. Asadel Publisher, Boulder
44. Xu C, Huang X, Zhu J, Zhang K (2018, January). Research on the construction of sanya smart tourism city based on internet and big data. In: 2018 International Conference on Intelligent Transportation, Big Data & Smart City (ICITBS). IEEE, pp 125–128
45. Zhong S, Di H (2017) Struggles with changing politics: street vendor livelihoods in contemporary China. Anthropological Considerations of Production, Exchange, Vending and Tourism. Emerald Publishing Limited, Leeds, England, pp 179–204
46. Zhou P (2024) SWOT Analysis of Economic Development in Baisha County, Hainan Province under the Background of the Free Trade Port. J Mod Bus Econ 1(2)
47. Zhou W (2022) The social shopping movement in China: from TaoBao to WeChat. In: The SAGE handbook of social media marketing. SAGE Publications Ltd., pp 205–220

Chapter 8
Innovation and Institutional Sustainability: ICT and Promoting the Local Tourism Industry

8.1 Introduction: The Role of ICT in Local Tourism Development

A digital revolution in destination marketing, service delivery, and visitor engagement has resulted from the swift development of information and communication technologies (ICTs), which have completely changed the global tourism sector [2, 30]. Digital platforms, mobile applications, and real-time online services have gradually supplanted traditional tourism promotion strategies, which were previously dominated by print ads, television commercials, and travel agencies [33]. In addition to redefining destination marketing, this shift has also changed governance frameworks, institutional structures, and stakeholder involvement in the growth of the tourism industry [9]. Because it provides creative, data-driven, and highly customized experiences that meet a range of travellers' needs, ICT-driven tourism promotion is now at the center of sustainable tourism strategies.

ICTs' capacity to democratize information access and promote face-to-face communication between service providers and tourists is among their most important contributions to tourism marketing. Destinations can interact with audiences around the world without the need for middlemen by using digital platforms like social media, search engines, and websites that offer travel reviews. Through user-generated content, dynamic pricing strategies, and AI-driven recommendations, platforms such as TripAdvisor, Booking.com, and Airbnb enable local communities and businesses to engage with tourists and showcase their unique offerings. Smaller travel destinations and independent tour operators now have equal visibility opportunities alongside well-known international brands thanks to this disintermediation.

Travelers' real-time interactions with destinations have been transformed by mobile applications and geolocation-based services, which go beyond marketing and visibility. In order to discover new locations, tourists now rely on augmented reality experiences, location-based recommendations, and smart maps. AI-driven suggestions based on user behaviour, live updates on local events, and customized

A. Cheshmehzangi, *Urban Innovation and Sustainability*, Urban Sustainability,
https://doi.org/10.1007/978-981-96-6370-5_8

itineraries are all provided by apps like Google Maps, Ctrip, and Yelp. Additionally, the efficiency of tourism services has been improved through the integration of digital ticketing, cashless payment systems, and smart booking platforms, making them more scalable, seamless, and accessible.

8.1.1 From ICT-Based Tourism to Promoting Local Tourism Industry

For ICT-based tourism promotion to benefit local economies and governance structures, institutional sustainability is essential [10, 43]. To support long-term industry growth, the shift to digital tourism necessitates robust regulatory frameworks, data protection laws, and infrastructure investments [23, 34]. ICT solutions have been incorporated into larger urban and regional development plans by numerous governments through the implementation of smart tourism policies [19, 20]. Maintaining a balance between quick digital adoption and environmentally friendly tourism management requires this institutional support. In order to build inclusive, resilient, and financially sustainable tourism ecosystems, government organizations, tech companies, and travel operators have come together to form public–private partnerships [7], which have become important catalysts for innovation in digital tourism.

But even though ICT has given local tourism businesses previously unheard-of opportunities, it has also brought about new difficulties and complicated governance. Significant obstacles to promoting equitable travel include the impact of algorithms on content visibility, the dangers of data monopolization, and the possibility of digital exclusion. In the algorithm-driven tourism economy, local companies and destinations that lack financial resources, technological infrastructure, or digital literacy frequently find it difficult to compete. Furthermore, since social media aesthetics frequently influence visitor expectations and occasionally misrepresent local cultures and environments, the rise of hyper-digitalized tourism experiences raises questions regarding the authenticity of travel narratives.

Institutional frameworks must change in response to these issues in order to facilitate inclusive digital adoption and maintain the sustainability, local focus, and cultural sensitivity of ICT-based tourism marketing. To develop digital policies that put fair access, community involvement, and responsible tourism growth first, governments, corporations, and communities must collaborate. Digital tourism promotion can be in line with more general urban sustainability objectives by utilizing ICT not only for financial benefits but also for cultural preservation, social inclusion, and environmental sustainability. This will increase the industry's resilience.

Strategic planning, the development of digital capacity, and cooperative governance models are necessary for ICT's integration into regional tourism economies as it continues to transform the travel industry. In addition to technological developments, the way institutions handle and control this shift is crucial to the success of ICT-based tourism promotion. In order to create a resilient, flexible, and socially inclusive tourism sector going forward, it will be essential to make sure that digital tourism is in line with long-term sustainability goals.

8.1.2 Digital Disruption in Tourism: The Rise of ICT-Based Promotion

ICT integration in tourism marketing has completely changed how businesses interact with their clientele, how travellers make decisions, and how destinations are advertised [8, 47, 49]. Because they allow for real-time, data-driven interaction with prospective tourists, digital platforms have emerged as the main tool used by travel agencies, hospitality providers, and destination marketing organizations (DMOs) worldwide [28]. By enhancing customer satisfaction, service delivery, and economic sustainability, smart tourism ecosystems—powered by blockchain, big data, and artificial intelligence (AI)—are revolutionizing the travel industry [39]. Even though digital innovation is changing the travel industry, institutional sustainability is still essential for long-term success.

The emergence of social media and influencer-driven travel marketing has been one of the biggest changes. Traditional brochures and TV ads have been supplanted by platforms like Instagram, TikTok, and YouTube, which enable travel destinations to create genuine and striking narratives [42]. Digital storytelling has been effectively used by campaigns like #VisitDubai and #PureNewZealand to present travel destinations as alluring, immersive, and experience-driven. Nevertheless, algorithm-driven content visibility poses difficulties because it is frequently difficult for smaller locations with fewer digital resources to become visible. Institutional assistance is needed to address these disparities through capacity-building initiatives, digital literacy training, and public–private partnerships that support new travel markets in maintaining their competitiveness in the digital sphere [14].

AI-powered chatbots and mobile applications are revolutionizing traveler engagement in addition to social media marketing. Travelers' convenience is increased by apps like Google Travel, Klook, and TripAdvisor, which offer real-time translations, personalized recommendations, and easy booking processes. AI is being used by smart tourism projects in places like Barcelona and Amsterdam to forecast visitor flows, improve crowd control, and strengthen the preservation of cultural heritage [27, 37]. By guiding travellers to lesser-known locations and relieving pressure on over-visited landmarks, AI-driven destination marketing also guarantees targeted outreach to various tourist segments, fostering sustainable tourism [29]. Regulating AI and big data use in tourism, protecting privacy, deploying AI ethically [35], and preventing data monopolization by large tech companies must be the main goals of institutional sustainability initiatives.

The use of immersive technologies like virtual reality (VR) and augmented reality (AR) is another significant advancement in the promotion of digital tourism [26].

For example, Japan's Kyoto VR experience lets visitors virtually explore cultural heritage sites, preserving historical landmarks while lessening the environmental impact of mass tourism [48]. By offering interactive virtual tours of museums and historical sites, Google's Arts & Culture platform expands accessibility worldwide [45]. To ensure that local companies and smaller tourism providers profit from these digital innovations as well, governments and tourism boards must create institutional policies that encourage these innovations while maintaining economic inclusivity.

Ensuring equitable access to ICT-driven opportunities is a major challenge in the field of digital tourism [6]. In order to fully benefit from online tourism promotion, many local tourism businesses—particularly those in developing nations—lack the digital infrastructure, funding, and technological know-how. To assist small-scale tourism operators in implementing ICT solutions, governments in nations like Thailand and Indonesia have implemented digital tourism grants and skill development initiatives. In order to ensure that digital transformation benefits all stakeholders—not just large travel companies or international booking platforms—these initiatives emphasize the importance of institutional sustainability measures.

Furthermore, blockchain technology is starting to revolutionize the field of digital tourism. Blockchain can improve data security, increase trust in peer-to-peer tourism services, and remove middlemen from travel bookings by offering transparent, safe, and decentralized transaction systems [15]. For instance, Malta's government has made it easier for travellers to apply for visas and increased security in the travel industry by implementing blockchain-based digital identity verification [15]. In order to ensure equitable access, moral data management, and regulatory compliance in digital tourism transactions, policymakers everywhere must take institutional frameworks for blockchain adoption into account.

Lastly, climate action and responsible tourism principles must be in line with sustainability in digital tourism promotion [44]. If not handled properly, an excessive dependence on digital marketing can lead to overtourism, cultural commodification, and environmental damage. Digital sustainability metrics are integrated into smart tourism governance models [1], like those in Norway and Costa Rica, to make sure that eco-friendly tourism policies and ICT-driven promotion are in line. To strike a balance between the expansion of tourism and environmental preservation, governments must implement institutional measures like visitor caps, ecotourism incentives, and digital carbon footprint tracking [5]. In summary, ICT's emergence in tourism marketing presents previously unheard-of chances for international travel destinations, but its long-term viability depends on robust institutional frameworks. To guarantee digital accessibility, data ethics, and economic inclusivity in the digital transformation of tourism, governments, corporations, and communities must cooperate. Tourism promotion can promote long-term economic resilience while maintaining cultural and environmental sustainability by implementing ethical AI deployment, responsible ICT policies, and equitable access to digital markets.

Fig. 8.1 Houhai Village and its beach areas have only become a major popular destination in recent years (The Author's Own)

8.2 Setting the Scene: Understanding Houhai Village's Tourism Landscape

8.2.1 The Case of Houhai Village: A Traditional Fishing Community Turned Tourism Hub

In Sanya, Hainan Province, China, Houhai Village, also called Tenghai Fishing Village, is situated next to the northern portion of Queens Bay at the southern end of Haitang Bay. In the past, Houhai was a tiny fishing community whose economy was mostly reliant on subsistence fishing and marine resources [32]. But over the last ten years, it has changed dramatically, becoming a popular surfing destination and ecotourism destination that attracts tourists from both domestic and foreign countries.

The emergence of ICT-driven tourism promotion and institutional sustainability measures has been one of the main forces behind this change. To increase the visibility of Houhai's tourism offerings, local tourism authorities have actively used digital platforms, e-commerce, and smart tourism strategies in partnership with businesses and community stakeholders. Today, the village is a popular destination for adventure tourism, cultural immersion, and sustainable travel experiences because of its surf-friendly beaches, relaxed coastal lifestyle, and eco-conscious development strategies (See Figs. 8.1, 8.2, 8.3, 8.4, 8.5, 8.6, 8.7, 8.8, and 8.9).

8.2.2 Houhai's Tourism Growth and the Role of ICT

The use of ICT in service delivery, visitor engagement, and destination promotion has greatly contributed to Houhai's explosive tourism growth (See Fig. 8.9). Unlike Sanya's major tourist hotspots, which rely heavily on large-scale luxury resorts and mass tourism strategies, Houhai has embraced digital innovation to develop a niche market for independent travellers, adventure seekers, and eco-conscious tourists.

The growth of influencer-driven travel and social media marketing has been crucial to this achievement. Houhai's online reputation as a lively and genuine surf tourism destination has been greatly influenced by digital platforms like WeChat, Douyin

Fig. 8.2 It took a lot of effort through online/digital tourism promotion to make Houhai a vibrant and popular destination for tourists (The Author's Own)

Fig. 8.3 Houhai continues to offer places of joy, excitement, and particular sports (such as surfing) to younger generations through multiple medias (The Author's Own)

Fig. 8.4 Surfing has resurfaced to become a major sport and leisure activity in Houhai Village (The Author's Own)

Fig. 8.5 There are now places to stay, take photo, have food, drink, spend a day, and spend the night across all Houhai Village (The Author's Own)

(TikTok's Chinese equivalent), Xiaohongshu (Little Red Book, or now known as RedNote), and Weibo. The village's local culture, beachside hostels, surf schools, and sustainable tourism experiences are highlighted in videos, blogs, and live streams regularly shared by surf enthusiasts and travel influencers. Houhai is now positioned

Fig. 8.6 All recent changes have led to Houhai becoming a hub for younger generations as well as those who enjoy sports, food, night life, and leisure activities at this cozy beachside town (The Author's Own)

Fig. 8.7 Fancy beach-side bars with their pools are often see extensively through social media platforms (The Author's Own)

Fig. 8.8 Social media marketing influences where people visit, stay and interact and how they choose their places of romance, gathering, parties, eating, drinking, etc. (The Author's Own)

as a growing tourist destination that is different from Sanya's popular tourist attractions [32] thanks to these digital narratives, which have greatly increased organic engagement with potential tourists.

Furthermore, Houhai's developing tourism ecosystem has benefited from smart tourism initiatives. To improve visitor convenience, local companies have implemented cashless payment methods, online reservation systems, and geolocation-based marketing. E-commerce platforms like Alipay and WeChat Pay, as well as online travel agencies (OTAs) like Ctrip and Meituan, have contributed to the growth of surf schools, homestays, and eco-friendly cafes by allowing small businesses to reach a wider audience without depending on conventional tourism middlemen.

8.2.3 Institutional Sustainability and Community-Led Tourism Development

Institutional support and community-led initiatives have also contributed to the tourism industry's sustainable growth in Houhai. By limiting overdevelopment, enforcing environmental regulations, and integrating smart infrastructure to support

Fig. 8.9 ICT at its best and worst in the same image. Mass tourism strategies, visitor engagement, service delivery are the backbone of ICT use in popular places like Houhai Village in Southern Hainan regions. Many people use major social platforms to decide where to go and stay while in Houhai. In fact, social media itself has given the chance to visitors and locals the opportunity to promote popular destinations. Tourism industry is growing fast in places like Houhai, and hence, digital innovation has extensively helped to develop a niche market for independent travellers, adventure seekers, and eco-conscious tourists. E-commerce platforms like Alipay and WeChat Pay, as well as online travel agencies (OTAs) like Ctrip and Meituan, have contributed to the growth of surf schools, homestays, and eco-friendly cafes (The Author's Own)

sustainable visitor experiences, the local government has actively promoted responsible tourism development in partnership with tourism boards and industry stakeholders.

For example, laws supporting low-impact tourism pursuits like kayaking, surfing, and cultural events have supported the village's economic resilience and ecological integrity. Through the operation of surf schools, guesthouses, seafood restaurants, and cultural workshops, Houhai's community-driven tourism model guarantees that locals actively engage in tourism entrepreneurship. Together with top-down institutional policies, this bottom-up strategy has produced a tourism framework that strikes a balance between local heritage preservation [21], environmental preservation, and economic growth.

Local small businesses and entrepreneurs now have more power thanks to digital technology, which enables them to reach larger markets and provide higher-quality services. In order to maximize local revenue generation and lessen reliance on big tourism corporations, local vendors can sell fresh seafood, advertise homestay experiences, and provide surfing lessons to tourists directly through social media campaigns, live streaming, and online marketplaces.

8.2.4 The Impact of ICT-Driven Strategies on Houhai's Tourism Expansion

The implementation of ICT-driven tourism strategies has had a profound impact on Houhai's economic sustainability and global visibility. Several key factors illustrate the role of digital innovation in its tourism success (Fig. 8.10):

1. ***Social Media and Influencer Marketing:***
    - o Surfing and lifestyle influencers regularly showcase Houhai as an alternative to mainstream beach destinations like Dadonghai and Yalong Bay.
    - o Travel blogs, vlogs, and live-streamed experiences generate authentic engagement, making it easier for potential visitors to discover the destination.
2. ***Smart Infrastructure and Digital Accessibility:***
    - o Mobile payment solutions and QR-code-based transactions have streamlined visitor spending, making it easier for tourists to book services, rent equipment, and support local businesses.
    - o Free Wi-Fi zones and tourism apps provide real-time travel information, including weather conditions, surf reports, and activity bookings.
3. ***E-Commerce and Online Travel Services:***
    - o Platforms like Meituan and Fliggy have enabled small businesses to market surf camps, eco-lodges, and local culinary experiences to a wider audience.
    - o The rise of digital tourism agencies has helped bridge the gap between independent travelers and local service providers, facilitating seamless trip planning.
4. ***Sustainable Tourism Innovation:***
    - o Digital platforms have been used to promote eco-friendly travel choices, such as electric scooter rentals, sustainable accommodation options, and carbon offset initiatives.

Fig. 8.10 Summary of key factors illustrating the role of digital innovation in its tourism success

- o ICT-enabled tourism management systems monitor visitor flow and environmental impact, ensuring long-term sustainability.

Houhai Village is a prime example of how ICT-driven tourism innovation can uphold sustainability principles while converting modest travel destinations into competitive tourism hotspots. In contrast to highly commercialized tourist destinations, Houhai's success is a result of its smart tourism infrastructure, community-led approach, and digital marketing tactics [22, 24]. A more resilient and inclusive tourism ecosystem has been made possible by local businesses' ability to use digital platforms for self-promotion, service delivery, and economic empowerment.

Institutional policies must continue to encourage digital adoption in the future while preserving Houhai's distinctive environmental and cultural resources. Houhai can firmly establish itself as a model for sustainable tourism development in China by enhancing digital accessibility, encouraging responsible tourism initiatives, and incorporating cutting-edge ICT solutions like blockchain-based visitor management and AI-driven analytics.

8.3 ICT-Driven Innovation in Houhai

8.3.1 ICT-Driven Strategies for Place Promotion in Houhai Village

Using digital storytelling, real-time visitor engagement, and smart tourism innovations, Houhai Village has adopted ICT-driven strategies to position itself as a sustainable and digitally connected tourism destination, drawing in an increasing number of both domestic and foreign visitors. Houhai's digital transformation, in contrast to conventional tourism promotion strategies, is driven by community involvement, entrepreneurial engagement, and institutional support, guaranteeing that technology not only increases visibility but also supports local economic sustainability.

The use of digital storytelling, mostly via social media sites like Weibo, Xiaohongshu (Little Red Book), and Douyin (TikTok), is a crucial component of Houhai's ICT-driven strategy. The village has been able to highlight its distinctive fusion of ecotourism activities, surfing culture, and the lively local way of life thanks to these platforms. Influencers, pro surfers, and travel vloggers are essential in promoting Houhai and establishing it as a relaxed substitute for Sanya's more marketed tourist destinations. Tourists can witness events like surf competitions, cultural performances, and local culinary showcases in real time through live-streamed content, which has significantly increased engagement from younger, energetic audiences keen to experience these genuine events.

The incorporation of smart tourism and real-time visitor engagement are also critical components of Houhai's digital transformation. AI-powered chatbots and WeChat Mini Programs are two examples of tools that provide individualized travel assistance, giving travellers access to the most recent data on eco-friendly travel options, weather forecasts, surfing conditions, and lodging availability. Furthermore, QR-code-based tourism experiences have grown in popularity because they offer visitors interactive content that showcases Houhai's cultural heritage, marine biodiversity, and local history. Applications for augmented reality (AR) further improve the experience of visitors by providing interactive maps that lead them through seafood markets, eco-tourist trails, and sustainable lodging, guaranteeing an educational and enjoyable trip.

Another important element in Houhai's digital transformation has been e-commerce. Tourists can now interact with local businesses, rent surfboards, reserve homestays, and eat at seafood markets more easily thanks to the widespread use of digital payment systems like Alipay and WeChat Pay. Additionally, by making lodging, surfing lessons, and guided tours easily accessible, online booking services like Ctrip, Meituan, and Fliggy have made trip planning easier. In line with the increasing demand for digital convenience in the travel industry, this e-commerce integration guarantees a smooth and convenient experience for guests. In order to interact directly with tourists and promote local economic growth, live commerce—especially on platforms like Douyin and Taobao Live—has created new opportunities for local business owners to sell goods like freshly caught seafood, eco-friendly travel packages, and handcrafted souvenirs.

Houhai Village has successfully closed the gap between digital marketing and the growth of sustainable tourism thanks to these ICT-driven innovations. These tactics not only make the village more visible as a distinctive travel destination, but they also guarantee that the financial gains from tourism are shared fairly while maintaining the village's natural and cultural integrity. Houhai is establishing a standard for how traditional communities can successfully incorporate ICT into tourism while upholding a sustainable and community-focused approach by embracing technology.

8.3.2 Social Media and Digital Storytelling for Cultural Branding

The development of Houhai Village as a sustainable tourism destination has been greatly aided by its clever use of digital storytelling and social media, which has been essential in establishing its cultural identity and highlighting its distinctive features. WeChat, Douyin (TikTok), Xiaohongshu (RedNote), and Weibo are just a few of the platforms that have helped make Houhai a genuine and interesting travel destination for contemporary travellers.

The foundation for interaction and communication with tourists has been WeChat. Houhai has established a direct line of communication with both prospective tourists and the general public through WeChat Mini Programs and official accounts. In addition to providing real-time updates on events, weather, and ecotourism options, these digital spaces showcase content that highlights regional customs like fishing methods and surf culture. The incorporation of payment systems such as WeChat Pay further streamlines transactions and gives travellers a seamless experience, making Houhai a practical and alluring option.

TikTok, also known as Douyin, has emerged as a major platform for user-generated content and influencer marketing. With its breathtaking beaches and active fishing community, Houhai's picturesque landscape has made it an ideal location for short-form video content. With videos that have millions of views, travel influencers and content producers highlight pursuits like surfing, regional cuisine, and cultural excursions. Viral hashtags and geotagging have raised awareness and increased visits, and younger, tech-savvy tourists find great resonance in these posts' genuineness.

Because it allows users to share personal stories and travel advice, Xiaohongshu, which is well-known for its lifestyle-focused content, is essential in forming Houhai's brand. Travelers share images, reviews, and suggestions for eco-friendly lodging choices, nearby companies, and eco-friendly events. These user-generated stories help to shape Houhai's growing reputation as a place that blends sustainability and cultural diversity.

Weibo further amplifies Houhai's digital presence by allowing the village to engage with both domestic and international audiences through real-time conversations, event promotions, and collaborations with influencers and local celebrities. Through these platforms, Houhai crafts a narrative that aligns with its values of eco-tourism, sustainability, and community engagement, ensuring that its cultural identity is promoted in a way that appeals to global audiences.

Social media and digital storytelling have thus enabled Houhai to rise above its traditional origins and become a destination that combines contemporary digital tactics with cultural preservation, ultimately establishing its standing as a cutting-edge, environmentally conscious, and culturally diverse tourism destination.

8.3.3 Smart Tourism Innovations: AI, Big Data, and Personalization

In order to improve the visitor experience and maximize tourism services, Houhai Village has embraced smart tourism innovations, utilizing state-of-the-art technologies such as big data, artificial intelligence, and geolocation-based services. These developments are essential to Houhai's development into a cutting-edge, eco-friendly travel destination that places an emphasis on individualized, data-driven services.

Big data analytics are essential for comprehending visitor trends and actions. Houhai can learn more about the tastes and routines of its tourists by examining the vast amounts of data produced by travellers via social media, smartphone apps, and reservation systems. This enables regional companies and travel agencies to customize their services according to current patterns, like the most popular times of year to visit, well-liked activities, or popular attractions. For example, information about beach traffic, weather predictions, and surf conditions is incorporated into digital platforms to help travellers make better decisions and enhance their trip.

Another essential component of Houhai's smart tourism approach is AI-driven suggestions. Based on a visitor's past visits, preferences, and even social media activity, the village's digital platforms can recommend tailored dining options, activities, and attractions using machine learning algorithms. For instance, the platform will suggest specialty eateries or surfing events if a user regularly engages with content about local seafood or surfing. This degree of customization encourages return visits and raises visitor satisfaction by resulting in a more interesting and tailored experience.

Promotions that use geolocation enhance the travel experience even more. Houhai can instantly send pertinent, location-specific offers to visitors' smartphones by utilizing GPS technology. Depending on how close they are to particular locations or businesses, visitors can get alerts about sales, special events, or nearby attractions. Tourists who are close to a local surf school, for instance, might be notified about a special class or discount, which would increase foot traffic and boost sales for nearby businesses. Collectively, these smart tourism innovations help to create a more individualized, responsive, and pleasurable travel experience in addition to optimizing the effectiveness of Houhai's tourism services. Houhai is establishing itself as a pioneer in eco-friendly, tech-driven travel by utilizing AI and big data.

8.4 Community-Led Digital Tourism and Economic Sustainability

The ICT-driven approach to tourism in Houhai Village has given local companies, small business owners, and community stakeholders a lot of chances to get involved in the tourism industry. Houhai has established a more sustainable and inclusive model of economic development by empowering its community to engage in and profit from the expanding tourism industry through the use of digital tools and online platforms.

The use of e-commerce solutions by nearby companies is among the most significant effects of this change. To advertise their services, make reservations, and sell goods, many of Houhai's small-scale vendors—such as surf schools, neighbourhood seafood markets, and craftspeople—have resorted to digital platforms. By exhibiting their products through captivating content, such as images, videos, and live-streams, platforms like WeChat, Douyin, and Xiaohongshu have enabled these companies to reach a wider audience both domestically and abroad. Along with helping local businesses become more visible, this has also assisted entrepreneurs in growing their clientele, which has resulted in higher sales and business expansion.

More effective reservation systems have also been made possible by the digital revolution, which is advantageous to both travellers and companies [36, 38]. Local accommodation providers, from guesthouses to boutique hotels, have integrated digital booking platforms that allow for easy reservations, direct communication with guests, and dynamic pricing strategies based on demand and seasonality. While providing visitors with a smooth and easy-to-use experience, this system assists businesses in optimizing occupancy rates.

Additionally, ICT integration has significantly improved the promotion of sustainable tourism businesses [12]. Through online channels, Houhai has promoted ecotourism, including sustainable seafood sourcing and responsible surfing tours, by utilizing its distinct cultural and environmental appeal. Houhai has drawn environmentally conscious tourists and helped local companies that place a high priority on sustainability by using digital tools to raise awareness of these eco-friendly initiatives. Houhai is positioned as a destination dedicated to responsible tourism thanks to digital storytelling platforms like social media, which enable community stakeholders to share the village's sustainability initiatives, from supporting local farmers to marine conservation.

In sum, Houhai's adoption of ICT has promoted sustainability and economic resilience by enabling local companies to engage with a global audience, adjust to digital trends, and advance sustainable practices. Small business owners and other stakeholders have been able to prosper in the tourism sector thanks to the community-led approach, which has also promoted sustainable development methods and produced financial gains [3, 41]. Local communities can use technology to improve economic sustainability while protecting their cultural and environmental heritage, as demonstrated by this digital tourism model.

8.4.1 Challenges and Institutional Barriers to ICT Adoption

Although Houhai Village's ICT-based tourism promotion strategy has produced notable benefits, a number of obstacles and institutional hurdles have prevented the region from realizing the full potential of digital transformation. The lack of infrastructure required to enable broad ICT adoption is one of the main challenges. Although Hainan's urban centers have seen large investments in digital infrastructure, rural and coastal areas like Houhai continue to struggle to access modern technologies and high-speed internet connectivity. Businesses and tourists alike are unable to fully take advantage of ICT innovations like real-time digital booking systems, intelligent tourism applications, and easy online payments because of these infrastructure gaps. Many local businesses, especially smaller ones, struggle with slow internet speeds and unreliable connectivity, which limits their ability to engage in the digital economy effectively.

The digital literacy gap, which impacts both local businesses and community members, is another important obstacle. Due to their inability to use digital tools and platforms, many small-scale vendors and business owners in Houhai find it challenging to manage e-commerce transactions, participate in online marketing, or even comprehend digital customer engagement tactics. A lack of technical training and knowledge separates those who can use ICT to grow their businesses from those who cannot, even with the availability of social media platforms and mobile applications. Achieving inclusive economic development is severely hampered by the digital divide, necessitating focused initiatives to raise community digital literacy.

Adoption of ICT in Houhai is also severely hampered by governance issues. The potential for ICT-based tourism development is constrained by the lack of clear policies or concerted efforts among local government agencies, tourism organizations, and the private sector. Effective promotion of Houhai as a sustainable tourism destination is hampered by the fragmented and inconsistent implementation of ICT-driven initiatives in the absence of a suitable governance framework. Lastly, one major worry is the possibility of over-commercialization [13]. The community's cultural and environmental heritage runs the risk of becoming commodified as the tourism sector grows and draws more tourists. Overtourism, a loss of local authenticity, and environmental damage could result from excessive commercialization fuelled by ICT-led marketing [31]. In order to maintain Houhai's distinctive character, stakeholders must strike a balance between digital promotion, sustainability, and community-led projects.

Interventions at the policy level are required to address these issues. These could include developing sustainable tourism governance frameworks, introducing regulations that reduce the risks of over-commercialization, enhancing digital infrastructure in rural areas, and putting in place digital literacy initiatives for nearby companies [16, 18]. To make sure that ICT adoption is in line with the long-term objectives of sustainable tourism growth and institutional resilience, cooperation between local government agencies, travel agencies, and the community will be crucial.

8.5 Future Directions: Strengthening Institutional Sustainability in Digital Tourism

The conclusions and conversations from the ICT-driven tourism promotion in Houhai Village show how promising digital transformation is for boosting the local tourism industry. Houhai must, however, scale digital tourism innovations and further strengthen institutional frameworks to maintain long-term sustainability [17]. Houhai's tourism industry's future depends on adopting new technologies, improving governance frameworks, and making sure that digital development prioritizes cultural and environmental preservation.

Using blockchain technology to improve the tourism value chain's efficiency, security, and transparency is one important future direction. Using blockchain technology, a transparent booking, payment, and transaction system can be established, increasing visitor and local business trust [4, 11]. Blockchain can assist in managing the environmental impact of tourism activities by enabling safe, real-time tracking of tourism data [25, 40]. This includes tracking local resource consumption and cutting waste in the tourism ecosystem.

The emergence of immersive virtual tourism is another fascinating area. With the help of technologies like virtual reality (VR) and augmented reality (AR), Houhai can provide prospective tourists with immersive experiences that let them learn about the village and its cultural offerings from a distance [32, 46]. As a creative marketing tool, virtual tourism can help Houhai reach a worldwide audience while lessening the strain on local resources and assisting in avoiding the dangers of overtourism. By offering digital displays of regional customs, celebrations, and historical locations without the need for physical intervention, virtual experiences can also help to preserve cultural heritage.

The implementation of smart governance frameworks will be essential to the expansion of digital tourism projects. Houhai's local government must create comprehensive, well-defined policies that govern the use of ICT to promote tourism while maintaining a focus on environmental and cultural sustainability. To create a well-rounded, forward-thinking tourism strategy that promotes responsible growth, cooperation between the local government, tourism stakeholders, and community leaders will be crucial. In order to ensure that the village can host tourists without placing an undue strain on the ecosystems or local infrastructure, smart governance can also involve the deployment of AI-powered systems to monitor and control visitor flows.

Enhancing digital access is still a key policy goal in addition to technology developments. A more inclusive digital economy can be achieved by ensuring that local companies and citizens have the resources, know-how, and tools necessary to interact with ICT platforms. In order to provide small business owners with the resources they need to leverage digital platforms for marketing, e-commerce, and customer interaction, governments and organizations ought to fund community-based digital literacy initiatives.

Lastly, a key component of Houhai's long-term tourism strategy will be striking a balance between economic expansion and environmental and cultural preservation. Digital tourism promotion needs to incorporate a more sustainable tourism model to prevent the dangers of over-commercialization and environmental damage. In conclusion, growing the digital ecosystem and coordinating ICT-driven innovations with institutional sustainability are key to the future of tourism in Houhai. A resilient and sustainable tourism model that benefits the community and the economy can be established in Houhai by embracing emerging technologies, bolstering governance, and guaranteeing equitable access to digital tools.

References

1. Abdelmalak F (2024) Smart tourism destinations: governance and resilience the use of ICTs in destination governance and its impact on resilience. Journal of Smart Tourism 4(2):35–45
2. Al-Adamat A (2015) The impact of information and communication technology on the marketing performance of Jordanian hotels. Doctoral dissertation, Queen Margaret University, Edinburgh
3. Armstrong R (2012) An analysis of the conditions for success of community based tourism enterprises. Available from: https://www.cabidigitallibrary.org/doi/full/10.5555/20123157768
4. Baralla G, Pinna A, Tonelli R, Marchesi M, Ibba S (2021) Ensuring transparency and traceability of food local products: a blockchain application to a Smart Tourism Region. Concurrency and Computation: Practice and Experience 33(1):e5857
5. Basu M (2024) Tech-driven horizons: pioneering sustainable tourism for environmental resilience and SDGs Triumph. In: The need for sustainable tourism in an era of global climate change: pathway to a greener future. Emerald Publishing Limited, pp 215–247
6. Bethapudi A (2013) The role of ICT in tourism industry. Journal of applied economics and business 1(4):67–79
7. Bordoloi K, Timung B, Das AM, Doloi G (2025) Digital platforms as catalysts for public-private partnerships in achieving sustainable development goals. Advancing sustainable development goals with educational technology. IGI Global Scientific Publishing, Hershey, Pennsylvania, pp 181–208
8. Buhalis D (2000) Tourism and information technologies: past, present and future. Tour Recreat Res 25(1):41–58
9. Caffyn A, Jobbins G (2003) Governance capacity and stakeholder interactions in the development and management of coastal tourism: examples from Morocco and Tunisia. J Sustain Tour 11(2–3):224–245
10. Çakar K (2023) Towards an ICT-led tourism governance: a systematic literature review. European Journal of Tourism Research 34:3404–3404

11. Calvaresi D, Leis M, Dubovitskaya A, Schegg R, Schumacher M (2019) Trust in tourism via blockchain technology: results from a systematic review. In: Information and Communication Technologies in Tourism 2019: Proceedings of the International Conference in Nicosia, Cyprus, January 30–February 1, 2019. Springer International Publishing, pp 304–317
12. Chakraborty PP (2024) The role of technology in enhancing sustainable tourism practices: innovations and impacts. In: Special interest trends for sustainable tourism. Hershey, Pennsylvania: IGI Global, pp 195–230
13. Cheshmehzangi A (2021) The significance of place in mind: comprehending memory through urban places. Urban Memory in City Transitions: The Significance of Place in Mind. Singapore: Springer, pp 61–119
14. Cheshmehzangi A (2022a) ICT for high levels of human capital in cities: a shift towards E-government and knowledge-driven economies. In: ICT, cities, and reaching positive peace. Singapore: Springer Nature, pp 139–157
15. Cheshmehzangi A (2022b) The application of ICT and smart technologies in cities and communities: An overview. ICT, cities, and reaching positive peace. Singapore: Springer Nature, pp 1–16
16. Cheshmehzangi A (2022) ICT, cities, and reaching positive peace. Springer, Singapore
17. Cheshmehzangi A (2022d) Exploring prospects to strengthen the nexus between ict, cities, and positive peace: limitations, challenges, and future directions. In: ICT, Cities, and Reaching Positive Peace. Singapore: Springer Nature Singapore, pp 219–235
18. Cheshmehzangi A (2023) Urban regeneration and the inevitable gentrification: the study of displaced communities in the city of Dali. In: Mapping urban regeneration: city life experiences in Yunnan, China. Singapore: Springer Nature, pp 77–104
19. Cheshmehzangi A, Dawodu A, Sharifi A (2021) Sustainable urbanism in China. Routledge, New York and London
20. Cheshmehzangi A, Tang T (2024) Navigating China's urban revolution: From tradition to transformation. China under construction: shaping cities through recent urban transformation. Springer, Singapore, pp 1–10
21. Cheshmehzangi A, Zou T, Tang T, Xie L (2023) Exploring influential factors for decision-making in urban development: urban heritage protection and displacing living communities. Bhumi, Plan Res J 10(1)
22. Dong E, Fu B, Li Y, Jin J, Hu H, Ma Y, Zhang Z, Xu Q, Cheng Z (2022) Hainan sport tourism development—a SWOT analysis. Sustainability 14(19):12653
23. Dredge D, Phi GTL, Mahadevan R, Meehan, E, Popescu E (2019) Digitalisation in Tourism: In-depth analysis of challenges and opportunities. Executive Agency for Small and Medium-sized Enterprises (EASME), European Commission. Available from: https://ec.europa.eu/docsroom/documents/33163/attachments/1/translations/en/renditions/native
24. Du S, Zhang H, Xu H, Yang J, Tu O (2019) To make the travel healthier: a new tourism personalized route recommendation algorithm. J Ambient Intell Humaniz Comput 10:3551–3562
25. Erol I, Neuhofer IO, Dogru T, Oztel A, Searcy C, Yorulmaz AC (2022) Improving sustainability in the tourism industry through blockchain technology: challenges and opportunities. Tour Manage 93:104628
26. Fan X, Jiang X, Deng N (2022) Immersive technology: a meta-analysis of augmented/virtual reality applications and their impact on tourism experience. Tour Manage 91:104534
27. Gokulapriyan R, Vinod L, Chandran RA, Yedla VR (2025) Smart city projects influencing smart tourist destinations. Advancing smart tourism through analytics. IGI Global Scientific Publishing, Hershey, Pennsylvania, pp 333–350
28. Gowreesunkar VG, Séraphin H, Morrison A (2018) Destination marketing organizations: Roles and challenges. In The Routledge handbook of destination marketing. Routledge, pp 16–34
29. Gün S (2025) Artificial intelligences impact on branding in tourism. Strategic brand management in the age of AI and Disruption. IGI Global Scientific Publishing, Hershey, Pennsylvania, pp 155–202

30. Hadjielias E, Christofi M, Christou P, Drotarova MH (2022) Digitalization, agility, and customer value in tourism. Technol Forecast Soc Chang 175:121334
31. Higgins-Desbiolles F, Carnicelli S, Krolikowski C, Wijesinghe G, Boluk K (2019) Degrowing tourism: rethinking tourism. J Sustain Tour 27(12):1926–1944, Available from: https://www.tandfonline.com/doi/shareview/10.1080/09669582.2019.1601732
32. Hu T, Chen H (2023) Identifying coastal cities from the perspective of "Identity-Structure-Meaning": a study of Urban tourism imagery in Sanya China. Sustainability 15(21):15365
33. Katsoni V, Spyriadis T (2020) Cultural and tourism innovation in the digital era. Springer International Publishing, Berlin/Heidelberg, Germany
34. Khater M, Ibrahim O, Sayed MNE, Faik M (2024) Legal frameworks for sustainable tourism: balancing environmental conservation and economic development. Current Issues in Tourism, 1–22
35. Law R, Ye H, Lei SSI (2025) Ethical artificial intelligence (AI): principles and practices. Int J Contemp Hosp Manag 37(1):279–295
36. Leung R (2022) Development of information and communication technology: from e-tourism to smart tourism. Handbook of e-Tourism, 23–55
37. Nag A (2024) Local development and tourism competitiveness: analyzing the economic impact of heritage tourism initiatives. In: Gastronomic Sustainability Solutions for Community and Tourism Resilience. Hershey, Pennsylvania: IGI Global, pp 160–190
38. Pencarelli T (2020) The digital revolution in the travel and tourism industry. Information technology & tourism 22(3):455–476
39. Pourroostaei Ardakani S, Cheshmehzangi A (2023) Big data analysis for an optimised classification for flight status: prediction analysis using machine learning classifiers. In: Big data analytics for smart transport and healthcare systems. Singapore: Springer Nature Singapore, pp 29–43
40. Rane N, Choudhary S, Rane J (2023, October 31) Sustainable tourism development using leading-edge artificial intelligence (AI), blockchain, internet of things (IoT), augmented reality (AR) and virtual reality (VR) technologies. Blockchain, Internet of Things (IoT), Augmented Reality (AR) and Virtual Reality (VR) technologies
41. Rytkönen S, Hotakainen S (2020) Promoting community-led conservation: opportunities, challenges and measures. Available online from: https://www.theseus.fi/handle/10024/349010
42. Serazio M (2023) The authenticity industries: keeping it "real" in media, culture, and politics. Stanford University Press, Redwood City
43. Shanker D (2008) ICT and tourism: challenges and opportunities, pp 50–58. Available from: http://115.249.96.25/xmlui/bitstream/handle/2259/185/50-58.pdf
44. Tehseen S, Hossain SM, Ong KY, Andrews E (2024) Sustainable tourism in a changing climate: balancing growth and environmental responsibility. The need for sustainable tourism in an era of global climate change: pathway to a greener future. Emerald Publishing Limited, Leeds, England, pp 69–94
45. Udell MK (2019) The museum of the infinite scroll: assessing the effectiveness of google arts and culture as a virtual tool for museum accessibility. Available online from: https://repository.usfca.edu/capstone/979/
46. Wang H, Zhang L, Yang F, Yan L, Lin C, Shen C (2025) Characteristics, source analysis, and risk assessment of organochlorine pesticide contamination in nearshore surface sediments of a tropical tourist island. Front Mar Sci 11:1513515
47. Werthner H, Klein S (1999) ICT and the changing landscape of global tourism distribution. Electron Mark 9(4):256–262
48. Yamada S (2017) Who moved my masterpiece? digital reproduction, replacement, and the vanishing cultural heritage of Kyoto. Int J Cult Prop 24(3):295–320
49. Zelenka J (2009) Information and communication technologies in tourism-influence, dynamics, trends. Informační Management, pp 123–132, Available from: https://dspace.zcu.cz/bitstreams/860a2ead-8994-4d2d-9c46-40d071731815/download

Chapter 9
Intersecting Paths: ICT's Role in Driving Innovation for a Sustainable Future

9.1 Reflections on ICT-Driven Innovation for Sustainability

ICTs' quick development has fundamentally changed how societies view sustainability by providing game-changing solutions in the institutional, social, economic, and environmental spheres. Digital innovations are now essential elements in building more inclusive, effective, and resilient communities rather than merely supporting tools [24]. Through digital governance, market modernization, social empowerment, and environmental conservation, this book has examined how ICT can help achieve sustainability and shown how these technologies can be effective facilitators of sustainable urban development.

The potential of ICT-driven innovation to support digital transformation in important industries is among its most significant effects. The incorporation of smart technologies, including blockchain, big data analytics, and artificial intelligence, has transformed decision-making procedures, maximizing resource allocation and fostering governance transparency. Urban management is now more data-driven thanks to digital platforms, which enable authorities to put in place real-time monitoring systems and evidence-based policies. ICT makes it possible for more efficient and scalable sustainability solutions, whether they are applied to waste management, energy efficiency, or public service delivery.

ICT promotes institutional sustainability from a governance standpoint by improving stakeholder participation, accountability, and efficiency [6]. Public administration has been enhanced by open-data platforms, e-government services, and digital governance frameworks, which have decreased bureaucratic inefficiencies and raised civic engagement. By offering digital platforms for citizen-led projects, public–private partnerships, and cross-sectoral collaboration, ICT also improves multi-stakeholder collaboration. Institutions can improve long-term sustainability by integrating technology into governance models, which will guarantee inclusive decision-making and promote more coherence in policy. ICT's contribution to community empowerment, especially in advancing social sustainability, is equally

A. Cheshmehzangi, *Urban Innovation and Sustainability*, Urban Sustainability,
https://doi.org/10.1007/978-981-96-6370-5_9

important. Digital tools give underprivileged and marginalized groups new access points to jobs, financial services, healthcare, and education [21, 32]. ICT has made economic participation more inclusive in urban settings, allowing small business owners to prosper through digital marketplaces and e-commerce platforms. More local representation in development processes is now possible thanks to the growth of social media, online engagement, and mobile technology, which has also improved community cohesion.

Additionally, ICT has opened up new avenues for economic sustainability by allowing companies to expand effectively while using the fewest resources possible. Real-time market analytics, intelligent logistics, and digital payment systems have improved supply chain management, cutting waste and guaranteeing more environmentally friendly corporate operations. ICT-driven solutions, like automated service platforms and digital tourism promotion, have increased economic opportunities while preserving environmental balance, illustrating the expanding relationship between responsible economic development and technological advancement. ICT will continue to play a key role in developing creative, inclusive, and robust solutions as the world navigates sustainability issues. Its ability to combine community development, governance, and economics guarantees that digital transformation is a fundamental component of sustainable futures rather than merely a change in technology.

The main ideas from the earlier chapters are summarized in this final chapter, which also considers how ICT-driven innovation contributes to sustainability in the social, institutional, economic, and environmental spheres. It talks about the wider ramifications for urban sustainability, policy-making, and upcoming ICT-driven strategies in addition to highlighting the lessons discovered from case studies in the Southern Hainan region, specifically in Sanya and the surrounding areas. The importance of interdisciplinary research and integrated practices is emphasized as the chapter also examines how these disparate approaches come together under a single vision for sustainable futures.

9.1.1 Converging Trends in ICT and (Urban) Sustainability

Since the early 2000s, the evolution of urban areas, economies, and governance structures is being redefined by the convergence of digitalization and sustainability. Through promoting inclusive governance, increasing economic efficiency, and optimizing urban management, digital technologies are becoming more and more important in boosting sustainability outcomes. Cities and communities can now more precisely and adaptably address sustainability challenges thanks to the broad adoption of ICT solutions, which range from digital platforms and smart infrastructure to AI-driven decision-making tools.

The cross-sectoral integration of ICT-based solutions in economic, governance, and urban management systems is one of the most important trends in this convergence [7, 13, 31]. Digital innovations are becoming more interconnected across various domains and are no longer limited to particular industries. Smart grids, IoT-enabled waste management, and AI-powered mobility planning, for instance, are revolutionizing resource allocation and service delivery in urban management [27]. In order to ensure more responsive and transparent decision-making, governance frameworks are increasingly integrating digital tools for e-governance, citizen engagement, and real-time data analytics [4, 30]. Through waste reduction, supply chain optimization, and inefficiencies reduction, economic systems—especially those based on automation, digital finance, and e-commerce—are becoming more sustainable [5, 23]. This systemic integration demonstrates how digital transformation is a deeply ingrained mechanism for attaining long-term sustainability rather than a stand-alone process.

Although technological advancements are driven by global ICT trends, meaningful sustainability outcomes depend on local and regional adaptation. The infrastructure, socioeconomic circumstances, and unique requirements of various communities must all be taken into account when implementing digitalization [14]. For example, while developed urban centers concentrate on AI-driven urban planning and smart energy solutions, emerging economies might place more emphasis on mobile technology and digital financial inclusion [9]. The implementation of ICT-based sustainability initiatives is also influenced by cultural, economic, and institutional differences, to optimize impact, localized governance models and adaptable policy frameworks are needed. In order to ensure that ICT acts as a catalyst for resilient and equitable urban futures, cities and regions must strike a balance between technological innovation and inclusive and adaptable approaches as digitalization continues to advance.

9.1.2 Case Study Synthesis: Insights from the Southern Hainan Region

This book's case studies, which highlight various aspects of sustainability in the Southern Hainan region, show how ICT-driven innovations are changing community and urban resilience. Each case provides important insights into the wider role of ICT in promoting sustainability by highlighting a particular area where digital technologies have been incorporated into economic development, governance, and environmental management.

Digital tools have been essential in monitoring coastal ecosystems, cutting waste, and putting smart governance strategies into practice in the context of beach management and environmental sustainability (see Chapter 4). In certain bays or beach areas, Sanya's coastal resilience has increased due to the integration of IoT-based monitoring systems, real-time data collection, and AI-driven environmental analytics, which guarantees that economic growth and tourism do not come at the expense of ecological deterioration. These developments show how digital solutions could potentially strike a balance between environmental stewardship and economic growth.

The example of smart residential compounds in Sanya, which addresses elderly communities and social sustainability (see Chapter 5), highlights how ICT can promote inclusive urban environments. Digital platforms have improved the quality of life for senior citizens by giving them access to community engagement tools, AI-driven help, and health monitoring services. IoT-based safety systems and digital healthcare services are integrated into smart residential planning, demonstrating how technology can be used to build age-friendly cities that put social inclusion and well-being first.

Through the use of social media, big data analytics, and clever marketing techniques, ICT has revolutionized the tourism sector, demonstrating the importance of place promotion and socio-economic sustainability (see Chapter 6). In Houhai Village and other parts of Sanya, platforms like WeChat, Douyin, and Xiaohongshu have completely changed the way that tourism is promoted, giving local entrepreneurs and small businesses a wider audience. Through revenue stream diversification and the creation of new opportunities for cultural tourism, the strategic application of influencer marketing and digital storytelling has improved economic resilience.

Traditional market ecosystems have been modernized through the integration of e-commerce, digital payment systems, and smart logistics within local markets and economic sustainability (see Chapter 7). Small businesses now have better access to finance because of the shift from cash-based transactions to mobile payments, which is evident in Sanya's main local markets. Online platforms have also helped vendors reach a wider audience, illustrating how digitalization promotes economic expansion while maintaining the integrity of regional trade.

Last, but not least, earlier discussions of institutional sustainability and tourism promotion emphasizes the importance of community-led digital tourism and ICT-enabled governance (see Chapter 8). Public–private partnerships that have aided in the development of digital infrastructure, training programs, and sustainable tourism regulations are linked to the success of Sanya's digital tourism strategies. These institutional initiatives demonstrate how digital governance frameworks can guarantee the tourism industry's long-term sustainability while empowering local communities.

When taken as a whole, these case studies (or case study samples) show that ICT is a catalyst for inclusive and resilient urban development in addition to being an efficient tool. The Southern Hainan region is a prime example of how digital transformation can promote social well-being, environmental preservation, and economic growth when it is in line with sustainability objectives.

9.2 Lessons Learned from ICT-Driven Sustainability Projects

This section identifies key takeaways from ICT-based sustainability projects in Sanya and beyond, shedding light on best practices, common challenges, and potential areas for improvement. Important insights into how ICT-driven innovations support sustainability can be gained from the case studies examined in and around Sanya. Digital technologies have transformed conventional methods and created new avenues for sustainable urban and community development in a number of areas, including environmental management, social well-being, economic development, and institutional governance. These projects provide insightful information for upcoming projects by highlighting both difficulties and best practices.

The value of flexibility in ICT implementation is among the most important lessons learned. Digital solutions must be customized to local contexts, making sure that technology is in line with cultural, economic, and environmental realities, as successful projects in Sanya have shown. Digital storytelling techniques on WeChat and Douyin, for example, have proven more successful in promoting tourism when combined with local narratives and community involvement. Similar to this, smart home solutions for senior communities have been successful when they have addressed particular needs, like AI-assisted communication tools and real-time health monitoring.

Another important lesson learned is that scaling ICT-based sustainability initiatives requires strong institutional support and governance. In order to ensure that investments in digital infrastructure are long-term viable, public–private partnerships have been essential to initiatives like smart beach management and digital market modernization. Coherent regulatory frameworks that promote digital innovation while addressing issues like data privacy, cybersecurity, and equitable access are still difficult to create, though. These governance concerns highlight the necessity of precise regulations that promote creativity while upholding moral and security principles. Economic inclusivity and financial sustainability also show up as major themes. Small businesses and vendors with low levels of digital literacy continue to face obstacles, even though ICT adoption has increased financial inclusion in local markets through e-commerce platforms and mobile payment systems [34]. Targeted training initiatives and easily navigable digital solutions that reduce entry barriers are needed to close this gap. Similarly, in order to avoid economic disparities, smart tourism initiatives must guarantee that digital transformation benefits both large corporations and small, locally owned businesses.

The digital divide is one issue that is frequently mentioned [19]. Disparities in digital literacy and infrastructure access are still noticeable, despite the fact that technological advancements have sped up the growth of Sanya's tourism, market, and residential sectors. Training expenditures, regional capacity-building initiatives, and infrastructure development in underprivileged areas are all necessary to guarantee digital equity. In the absence of these steps, the advantages of digitization run the risk of being concentrated among a small number of people rather than supporting more general sustainability objectives. Lastly, the case studies highlight the dynamic nature of ICT-driven sustainability. For digital strategies to be improved over time, stakeholder engagement, data-driven optimization, and ongoing assessment are crucial. Long-term success in any field—economic revitalization, social services, or environmental monitoring—requires a dedication to inclusive policymaking and iterative learning that synchronizes digital advancement with sustainability goals.

9.2.1 Effective Strategies for ICT-Enabled Sustainability

ICT-driven sustainability projects need to be incorporated into robust local governance frameworks in order to have a significant impact. Municipal and regional governments are essential in incorporating digital technologies into policy frameworks and making sure that ICT solutions complement long-term sustainability objectives. In addition to infrastructure investment, effective governance necessitates regulatory policies that promote digital innovation while preserving cybersecurity, data privacy, and fair access [15]. Even the most cutting-edge technological solutions run the risk of eventually becoming fragmented or unsustainable in the absence of institutional commitment.

Engaging multiple stakeholders is another crucial tactic for sustainability enabled by ICT. Governments, the private sector, academic institutions, and local communities must work together to achieve sustainable urban planning and digital transformation. Though their efforts must be directed by local needs and public sector regulations, the private sector—especially technology companies and digital service providers—can spur innovation. In order to guarantee that ICT solutions meet actual social and economic needs, community organizations and local business owners contribute crucial firsthand viewpoints. When it comes to supporting digital literacy initiatives, financing smart infrastructure, and jointly creating solutions that benefit a wide range of stakeholders, public–private partnerships are especially successful.

To guarantee that the advantages of sustainability are felt by all facets of society, technological innovation must be balanced with community-driven strategies. ICTs (including ICT-driven/ICT-based interventions or innovations) must be tailored to local socioeconomic circumstances and cultural contexts, even though it offers efficiency and scalability. The efficacy of ICT interventions is increased by bottom-up digital transformation techniques like community-led digital platforms, localized app development, and participatory planning. Technology should enhance and magnify local knowledge rather than replace it, promoting inclusive growth. In addition to enhancing governance and economic resilience, ICT solutions created with and for communities also contribute to the development of a more sustainable and adaptable urban future.

9.2.2 Challenges and Limitations of ICT Adoption

ICT has the ability to revolutionize sustainability, but there are still large infrastructure gaps and worries about the digital divide. Widespread ICT adoption is hampered by disparities in technological literacy, antiquated digital infrastructure, and unequal access to high-speed internet. Socioeconomic disparities are exacerbated in rural and underserved communities, which frequently lack the resources and connectivity necessary to fully benefit from digital transformation. ICT-driven sustainability initiatives run the risk of escalating rather than reducing current disparities if focused interventions are not made to close this digital divide [3, 8]. To guarantee that no community is left behind, governments and private sector stakeholders must place a high priority on funding inclusive capacity-building programs and equitable digital infrastructure.

Over-commercialization and overtourism are risks associated with the fast digitization of urban economies and tourism. Unchecked digital promotion can result in an unsustainable influx of tourists, overwhelming local resources, straining environmental ecosystems, and uprooting local communities, even though ICT facilitates effective marketing and visitor engagement. Finding a balance between long-term environmental and cultural preservation and economic gains is a challenge in places where ICT-driven place promotion has spurred exponential tourism growth. Digital tourism campaigns have the potential to unintentionally accelerate ecological degradation and socio-cultural disruptions in the absence of strategic regulations and intelligent management tools, thereby undermining the sustainability they are intended to promote.

ICT integration into sustainability initiatives is further hampered by regulatory and policy bottlenecks. Outdated governance systems that are sluggish to adopt digital innovations plague many cities and regions. Innovation is stifled by bureaucratic inefficiencies, strict policy frameworks, and ambiguous laws pertaining to platform governance, cybersecurity, and data privacy. Businesses and communities struggle to use ICT for sustainable growth in the absence of clear, progressive regulatory guidelines. In order to promote responsible ICT-driven development and protect social and environmental interests, governments must proactively streamline digital governance, create flexible policies, and cultivate regulatory environments.

9.3 Future Pathways: ICT and the Next Frontier of Urban Sustainability

As interdisciplinary approaches and emerging technologies come together to create more resilient, efficient, and inclusive cities, the future of urban sustainability rests on the ongoing development of ICT-driven innovation. Leveraging ICT will be crucial for improving governance, promoting more sustainable economic models, and optimizing resource management as urbanization picks up speed. The incorporation of AI and machine learning into urban systems will be among the most important developments. Cities can anticipate and address issues with energy use, mobility, and environmental management with the help of predictive analytics. Policymakers will be able to automate administrative tasks, improve public service delivery, and create data-informed policies thanks to AI-driven insights. Real-time data can be aggregated by smart city platforms, offering a proactive and adaptive dynamic approach to urban planning.

Urban experiences will be redefined by immersive technologies like virtual reality (VR) and augmented reality (AR) [16], especially in the areas of heritage conservation and smart tourism. These resources can facilitate virtual travel, offer interactive cultural experiences, and aid in the preservation of cultural heritage through digital storytelling. AR and VR can develop sustainable tourism models that lessen the physical burden on historic sites while increasing access to a variety of audiences by bridging the gap between digital innovation and cultural heritage. Cross-sector cooperation must be given top priority in future ICT adoption pathways for sustainability in order to guarantee that technological developments complement environmental and social justice objectives. ICT has the potential to propel a genuinely transformative and sustainable urban future by fusing cutting-edge digital solutions with economic strategies, governance frameworks, and community-driven projects.

9.3.1 Emerging Technologies for Sustainable Development

As interdisciplinary approaches and emerging technologies come together to create more resilient, efficient, and inclusive cities, the future of urban sustainability rests on the ongoing development of ICT-driven innovation. Leveraging ICT will be crucial for improving governance, promoting more sustainable economic models, and optimizing resource management as urbanization picks up speed.

The incorporation of AI and machine learning into urban systems will be among the most important developments [1, 22, 25, 33]. Cities can anticipate and address issues with energy use, mobility, and environmental management with the help of predictive analytics. Policymakers will be able to automate administrative tasks, improve public service delivery, and create data-informed policies thanks to AI-driven insights. Real-time data can be aggregated by smart city platforms, offering a proactive and adaptive dynamic approach to urban planning. As highlighted earlier, urban experiences will be redefined by immersive technologies like virtual reality (VR) and augmented reality (AR), especially in the areas of heritage conservation and smart tourism. These resources can facilitate virtual travel, offer interactive cultural experiences, and aid in the preservation of cultural heritage through digital storytelling. AR and VR can develop sustainable tourism models [26] that lessen the physical burden on historic sites while increasing access to a variety of audiences by bridging the gap between digital innovation and cultural heritage.

By improving governance, transparency, and operational efficiency, emerging technologies have the potential to further revolutionize sustainable development. By lowering fraud and facilitating real-time tracking of goods and services, blockchain applications can guarantee ethical supply chain management and transparent governance. Resource allocation will continue to be optimized by AI and machine learning, which will offer data-driven insights for waste management, energy conservation, and intelligent infrastructure planning. In the meantime, immersive technologies will be essential to sustainable tourism since they let tourists interact with the natural and cultural heritage in ways that have the least negative effects on the environment. These developments highlight the necessity of comprehensive ICT integration that supports sustainability objectives in order to provide urban communities with long-term social, economic, and environmental advantages.

9.3.2 *Building Institutional Resilience Through Digital Transformation*

Building institutional resilience through digital transformation is essential for developing strong and flexible governance structures as cities and regions continue to navigate the challenges of urbanization and sustainability. The first step in incorporating ICT solutions into urban planning is to fortify the legislative frameworks that facilitate digital sustainability governance. The intricacies of digital transformation must be addressed by effective policies that strike a balance between technological advancement and sustainable development objectives. These frameworks ought to offer protections to guarantee social justice, environmental preservation, and digital adoption in addition to incentives.

Promoting collaborations across sectors is equally important. To promote innovation, exchange knowledge, and make use of combined resources, governments, private businesses, and research institutes must collaborate [20]. In order to solve complicated urban sustainability issues, whether through more inclusive economic models, smarter infrastructure, or more effective energy systems, this collaboration is crucial. Public–private partnerships can provide access to capital and knowledge, enabling the creation of ICT solutions that are both scalable and responsive to regional demands.

Furthermore, to guarantee long-term digital inclusion, local capacity-building initiatives must be strengthened. Local stakeholders can take an active role in the digital transformation process by funding training programs, digital literacy initiatives, and community engagement. In addition to increasing social equity, enabling communities to fully utilize ICT lays the groundwork for resilient and inclusive sustainable development. By enhancing these capabilities, local institutions will be able to adjust to emerging technologies and make sure that digital transformation plays a significant role in future sustainability goals [2]. In conclusion, developing sustainable, future-ready cities requires a strategic digital transformation to strengthen institutional resilience. Institutions can be equipped to handle a world that is becoming more complicated and interconnected by integrating policies, encouraging partnerships, and developing local capacity.

9.4 The Road Ahead: Integrating ICT and Sustainability for Future Urban Success

It is obvious that ICT will become more and more important in determining urban sustainability worldwide as we move forward. When used strategically, ICT's transformative potential can open up new avenues for long-term urban resilience, promoting social and economic advancement in addition to environmental sustainability. It has already been demonstrated that integrating ICT into urban systems can maximize resource management, strengthen governance, improve public services, and encourage inclusive growth. But as this book shows, ICT needs to be governed by the values of environmental stewardship, equity, and inclusivity if it is to actually contribute to sustainable urban futures.

From sustainable tourism and market modernization to community empowerment, the case studies from the Southern Hainan region—especially the Sanya initiatives—provide a powerful example of how digital transformation can revitalize urban systems. These illustrations highlight the necessity of constant innovation and adaptation to make sure that ICT interventions are suited to regional contexts and in line with more general sustainability objectives [10]. The insights gained from these cases provide important information about how cutting-edge technologies like blockchain, artificial intelligence, and immersive tools can promote more intelligent and sustainable urban settings [17, 18, 28, 29].

The recommendations are clear for researchers, policymakers, urban planners, and industry stakeholders: ICT-driven sustainability requires a comprehensive, team-based approach. While urban planners should concentrate on developing infrastructure that seamlessly integrates digital solutions, policymakers should fortify digital governance frameworks and promote cross-sectoral collaborations. The potential of emerging technologies to tackle the intricate problems of urban sustainability must be further investigated by researchers. Stakeholders in the industry play a crucial part [11] in making sure that innovations are affordable, accessible, and inclusive.

As we move forward, integrating ICT into the fabric of urban development will be key to building resilient, sustainable cities. By embracing this digital transformation with foresight and responsibility, we can lay the foundation for urban futures that are not only technologically advanced but also socially and environmentally sustainable [12]. Building resilient, sustainable cities in the future will depend on incorporating ICT into the urban development process. We can create the groundwork for technologically sophisticated, socially and environmentally sustainable urban futures by embracing this digital revolution with responsibility and foresight. From resource optimization to inclusive growth, the nexus of cutting-edge technologies and sustainability presents an unmatched chance to tackle intricate urban issues. ICT solutions must be developed to promote long-term resilience as urban populations continue to rise. This will guarantee that cities continue to be flexible, dynamic, and able to serve the needs of their residents. In the end, ICT and sustainability practices working together will improve urban life quality and pave the way for a more sustainable future, which includes a variety of aspects in understanding and achieving urban sustainability.

References

1. Abbas M, Akhai S, Abbas U, Jafri R, Arif SM (2025) AI-enabled sustainable urban planning and management. In: Real-World Applications of AI Innovation. IGI Global Scientific Publishing, pp 233–260
2. Agboola OP, Tunay M (2023) Urban resilience in the digital age: The influence of Information-Communication Technology for sustainability. J Clean Prod 428:139304
3. Ali AH (2011) The power of social media in developing nations: New tools for closing the global digital divide and beyond. Harv. Hum. Rts. J. 24:185
4. Ali M (2023) E-governance and E-democracy: a Digital Revolution. Available at SSRN 4623414
5. Andrei N, Scarlat C, Ioanid A (2024) Transforming E-commerce logistics: sustainable practices through autonomous maritime and last-mile transportation solutions. Logistics 8(3):71
6. Bäckstrand K (2006) Multi-stakeholder partnerships for sustainable development: rethinking legitimacy, accountability and effectiveness. Eur Environ 16(5):290–306
7. Bajpai N, Beriya A, Biberman J, Sharma A, Wadhwa M (2019) Intersectoral Convergence and ICTs: Integrated ICT Approaches to Rural Poverty Reduction, No 10. ICT India Working Paper
8. Bandyopadhyay S, Bardhan A, Dey P, Bhattacharyya S (2021) Bridging the education divide using social technologies. Springer, Singapore
9. Ben Dhaou S, Isagah T, Distor C, Ruas I (2024) Global assessment of responsible artificial intelligence in cities: research and recommendations to leverage AI for people-centred smart cities. Available online from: https://collections.unu.edu/view/UNU:9789
10. Braa J, Monteiro E, Sahay S (2004) Networks of action: sustainable health information systems across developing countries. MIS quarterly:337–362
11. Chen W, Zhang S, Kong D, Zou T, Zhang Y, Cheshmehzangi A (2023) Spatio-temporal evolution and influencing factors of China's ICT service industry. Sci Rep 13(1):9703
12. Cheshmehzangi A (2021) From transitions to transformation: a brief review of the potential impacts of COVID-19 on boosting digitization, digitalization, and systems thinking in the built environment. Journal of Building Construction and Planning Research 9(1):26–39
13. Cheshmehzangi A (2022) ICT, cities, and reaching positive peace. Springer, Singapore
14. Cheshmehzangi A (2022) The impacts of ICT platforms on public place use in China. Built Environment 48(2):264–279
15. Cheshmehzangi A (2022c) The application of ICT and smart technologies in cities and communities: An overview. ICT, Cities, and Reaching Positive Peace, 1–16
16. Cheshmehzangi A, Ornsby P (2017) Augmented locality: the utilisation of urban screens in public places as new networks of the city. Journal of Urban Regeneration & Renewal 11(2):194–211
17. Cheshmehzangi A, Su Z, Zou T (2023) ICT applications and the COVID-19 pandemic: Impacts on the individual's digital data, digital privacy, and data protection. Frontiers in human dynamics 5:971504
18. Cheshmehzangi A, Batty M, Allam Z, Jones DS (2024a) Conclusions and the Future of City Information Modelling (CIM). In City Information Modelling. Singapore: Springer Nature, pp 249–253

19. Cheshmehzangi A, Zou T, Su Z, Dawodu A, Ardakani SP, Sedrez M, Tang T (2024) Growing digital divide for expatriates population in China: a user perspective analysis during the COVID-19 Pandemic. SAGE Open 14(4):21582440241274612
20. Etzkowitz H (2003) Innovation in innovation: The triple helix of university-industry-government relations. Soc Sci Inf 42(3):293–337
21. Eunice, O. N., Oluwatoyin, O. M., Tajudeen, A. O., & Lois, O. O. (2024). The impact of Digital Literacy on poverty alleviation: A psychosocial perspective. *Educational Perspectives, 12*(3).
22. Heidari A, Navimipour NJ, Unal M (2022) Applications of ML/DL in the management of smart cities and societies based on new trends in information technologies: A systematic literature review. Sustain Cities Soc 85:104089
23. Kaul D, Khurana R (2022) Ai-driven optimization models for e-commerce supply chain operations: Demand prediction, inventory management, and delivery time reduction with cost efficiency considerations. International Journal of Social Analytics 7(12):59–77
24. Misuraca G, Pasi G, Viscusi G (2018, April) Social innovation and resilience: Exploring the dynamics and impacts on the digital transformation of governance & society. In Proceedings of the 11th international conference on theory and practice of electronic governance, pp 91–100
25. Ogundare E (2024) Understanding the mediating role of artificial intelligence in urban transportation planning for smart city development and its implications for the United States. International Journal of Innovative Science and Research Technology 9:10–5281
26. Samaddar K, & Mondal S (2024) AR and VR-based travel: A responsible practice towards sustainable tourism. International Journal of Tourism Cities 10(1):105–128
27. Shafik W (2024) Incorporating Artificial Intelligence for Urban and Smart Cities' Sustainability. In: Maintaining a Sustainable World in the Nexus of Environmental Science and AI. IGI Global, pp 23–58
28. Pourroostaei Ardakani S, Cheshmehzangi A (2023a) Big data analytics and the future of smart transport and healthcare systems. In: Big Data Analytics for Smart Transport and Healthcare Systems. Singapore: Springer, pp 175–184
29. Pourroostaei S, Cheshmehzangi A (2023) Big data analytics for smart urban systems. Springer, Singapore
30. Troitiño DR, Mazur V, Kerikmäe T (2024) E-governance and integration in the European union. Internet of Things 27:101321
31. Viana LBF (2025) Redefining collaborative governance in the digital age: insights from an integrative systematic review on the role of ICT in collaboration. Technium Soc Sci J 67:303
32. Wu J, Guo S, Huang H, Liu W, Xiang Y (2018) Information and communications technologies for sustainable development goals: state-of-the-art, needs and perspectives. IEEE Communications Surveys & Tutorials 20(3):2389–2406
33. Yan Z, Jiang L, Huang X, Zhang L, Zhou X (2023) Intelligent urbanism with artificial intelligence in shaping tomorrow's smart cities: current developments, trends, and future directions. Journal of Cloud Computing 12(1):179
34. Zou T, Cheshmehzangi A (2022) ICT adoption and booming e-commerce usage in the COVID-19 era. Front Psychol 13:916843